Raspberry Pi 4 2020 BEGINNERS Guide (LARGE PRINT EDITION)

A Complete 2020 Manual to get started with Raspberry pi 4 Projects

TED HUMPHREY

Copyright

Copyright©2020 Ted Humphrey

ISBN: 9798667711728

All rights reserved. No part of this book may be reproduced or used in any manner without the prior written permission of the copyright owner, except for the use of brief quotations in a book review.

While the advice and information in this book are believed to be true and accurate at the date of publication, neither the authors nor the editors nor the publisher can accept any legal responsibility for any errors or omissions that may be made. The publisher makes no warranty, express or implied, with respect to the material contained herein.

Printed on acid-free paper.

Contents

Copyright i
CHAPTER ONE 1
INTRODUCTION TO RASPBERRY PI 4 1
The Raspberry Generations 2
Operating systems used by the Raspberry pi 7
Installing Raspbian on Raspberry pi 4 using NOOBS 8
Installing Raspberry on the Raspberry pi 4 by writing a disc image 11
Getting Familiar with the Raspberry Pi 4 14
Exploring the Raspbian desktop 17
Using your Raspberry Pi 4 19
Browsing the internet with the Raspberry pi 4: The chromium web browser 19
Saving your files on the Raspberry pi: The File Manager 21
Writing on the Raspberry Pi 4: The LibreOffice Productivity Suite 24
The Recommended Software tools 26
The Raspberry pi configuration tools 27
Shutting down 28
CHAPTER TWO 29
UNDERSTANDING THE RASPBERRY COMMAND LINE 29
Command for listing and accessing files 29

Installing software with the command line 30

Personalizing the command lines 33

Command lines for managing removable disc 34

Command line for network 34

Command lines to stop an ongoing process... 35

Command line to put a program in background 36

Accessing the Raspberry pi remotely from other PCs using SSH 37

CHAPTER THREE 38

THE RASPBERRY PROJECTS: BUILDING WITH THE Pi..38

The Mathematica Telescope 38

Poopi and piter's paludarium project... 39

Digital Zoetrope 39

Internet of Things Chessboard 40

Naturebytes wildlife cam kit 41

LIFEBOX 41

JOYTONE 42

PiScan.. 42

Watch iPlayer on Raspberry pi 43

Shooting in Slo-Mo with the Raspberry pi camera module 46

CHAPTER FOUR 52

PYTHON PROGRAMMING FOR RASPBERRY 52

Understanding the python environment... 53

The IDLE development environment shell 56

Running a python script in IDLE 58

Creating a python script in IDLE 59

Understanding the Python Basics 60

Producing python script output 60

Output control with Escape sequences 62

Writing comments in scripts 64

Understanding the python variable 65

Assigning value to a variable 67

Formatting variable and string output 67

Assigning long string values to variables 67

Assigning numeric values to variables 69

Reassigning a variable 69

Python data type 70

Allowing python script input 71

Working with numbers and performing calculations in the Raspberry python script 74

Python Math operators 74

Order of operation 75

Using variables in calculation 75

Creating complex numbers in python 77

Using the NumPy Math Libraries 78

Creating NumPy Arrays 80

iv

Learning about Loop in python 85

Creating and using Tuple in programming 92

Creating and using lists in python 93

CHAPTER FIVE **95**

USING PROGRAMMING IN THE RASPBERRY PI **95**

The Graphical User Interface (GUI) Programming 95

The tkinter package 98

Exploring the tkinter widget 108

Python Web Programming 109

Running a web server on your pi 109

CHAPTER SIX **112**

Basic pi/python projects **112**

Building your pi projects with python 112

ABOUT AUTHOR **127**

CHAPTER ONE

INTRODUCTION TO RASPBERRY PI 4

How easier can it be to have a computer as small as the size of a credit card doing for us what most bigger and sophisticated computers can do while utilizing a lower power unit? The Raspberry pi is just like a computer where users can get to design and customize its interface for many user-defined functions and projects. Much like your computer, the Raspberry pi requires an operating system that enables it to carry out a number of tasks just like your everyday computers. The Raspberry pi models earlier designed were not intended to gain much popularity, but the little single-board computers were functioning and were accepted by programmers, students and collaborators all over the world. The popularity was born accredited to the fact that the Raspberry pi series were among the first computers of its kind to allow users to design what they want to use it for- ranging from programming, robotics, creating other physical devices, browse the web and play a number of sophisticated video games amidst other interesting uses. One thing about the Raspberry pi, just like any other computer, is that much hardware can be attached to it through the General Purpose Input-Output Pins (GPIO) attached to the side of the Raspberry pi. But unlike most computers, the hardware attached to the Raspberry pi can be used for a number of different purposes such as temperature control hardware, LEDs, switches and others such as robots, motion detection gadgets, drones etc. NASA once used the Raspberry pi 3, one of the generations of Raspberry pi, to power their open source Rover – an indication that the Raspberry computers are applicable in space exploration.

The first model of the Raspberry pi, which is the Raspberry pi Model B launched in 2012, came with lesser features when compared to other models released afterwards. A good example is the Raspberry pi zero series which is a small and improved version of the much bigger sized Raspberry pi. The pi zero series came with better features such as the many USB ports and low power usage. Moving from the Raspberry zero series to the latest pi 4, the Raspberry pi series were observed to have a common feature which is that software for the pi zero can work with that of pi 4 – only that it is slower.

The Raspberry Generations

Since 2012, when the first model of the Raspberry pi was released, the Raspberry inventors did not stop giving their users many extraordinary improvements, which are usually incorporated into the newest versions of the Raspberry. But the common thing about all the versions of Raspberry released till date is that they all feature; **system on a chip (SoC)** together with a powerfully integrated **ARM**-packed **CPU** and the **on-chip graphics processing unit (GPU).** The **system on a chip (SoC)** was designed by the Broadcom Company, which is the leading designer of multi-functional semiconductor devices and other software. The **ARM** is a reduced instruction set computing architecture (RISC) primarily made for computer processors. The **on-chip graphics processing unit (GPU)** is responsible, primarily, for image creation ability of the Raspberry pi. There is a **secured digital (**SD) card which stores the Raspberry's operating system and all of its program memory.

- **Raspberry pi Model B:** this is the first model of the Raspberry series that came into the market in 2012. The **Raspberry pi model B+,** featuring an improved board design compared to the **pi model B** was released into the market in 2014.
- **Raspberry pi 2:** housed a 900 MHz ARM processor with 1 GiB RAM. The GiB (gibibyte) here differs slightly from GB (gigabyte). The pi 2 came into the market in February 2015. The earlier pi 2 model, which was the V1.1 model of the Raspberry Pi 2, utilized a Broadcom **BCM2836 SoC** with a 900 MHz 32-bit, ARM processor, with and a 256 KiB shared L2 cache. The later model, which was the Raspberry Pi 2 V1.2, was advanced to a Broadcom **BCM2837 SoC** featuring a 1.2 GHz 64-bit ARM A53 processor, the same SoC which is used on the Raspberry Pi 3, but underclocked (by default) to the exact 900 MHz CPU clock speed as the V1.1. The BCM2836 SoC went defunct as of late 2016.

- **The Raspberry pi zero:** The Raspberry pi zero was not exactly bigger as compared to the Raspberry pi 2 series. The pi zero also featured less input and output capabilities when compared with the Raspberry pi 2 series. The pi zero was released in November 2015. By February 2017, the company launched the **pi zero W** was released into the market, which featured a Wi-Fi and Bluetooth mode. About 10 months later, the **pi zero WH** was released into the market featuring a GPIO header. Both the Raspberry Pi Zero and Zero W feature 512 MiB of RAM. The pi zero, lacking any Ethernet circuit, is usually connected using an external Wi-Fi Adapter. A thorough look at the Pi Zero observed that the USB port is connected straight to the SoC, and that it deploys a micro USB (OTG) port. When compared to most other pi versions, the 40 pin GPIO connector is not available on the Pi Zero but the latter Pi Zero WH solves this.

- **Raspberry pi 3 model B and B+:** The pi 3 model B was launched into the market in February 2016. It featured a 1.2GHz 64 bit quad-core processor, a superb on-board 802.11n Wi-Fi, Bluetooth and USB boot modes. In 2018, the **Raspberry Pi 3 Model B+** was released which featured a faster 1.4 GHz processor and an Ethernet or 2.4 / 5 GHz dual-band 802.11ac Wi-Fi (100 Mbit/s). It also came with a Power over Ethernet (PoE) (with the add-on Power over Ethernet HAT), USB boot and network boot (which makes an SD card a defunct feature in the pi model b+). The Raspberry Pi 3 Model B utilizes a Broadcom BCM2837 SoC featuring a 1.2 GHz 64-bit ARM A53 processor, with a 512 KiB shared L2 cache. The Model A+ and B+ are both 1.4 GHz ARM processors.

- **Raspberry pi 4 model B:** Released in the first half of 2019 featuring a 1.2GHz 64 bit quad-core processor, a superb on-board 802.11n Wi-Fi, Bluetooth 5, two USB 2.0 ports and two USB 3.0 ports, and a Micro HDMI port supporting a dual monitor display showing a 4k resolution. The Raspberry Pi 4 utilizes a Broadcom BCM2711 SoC featuring a 1.5 GHz 64-

bit ARM A72 processor, with a 1MiB shared L2 cache. The Raspberry Pi 4 is available with 2, 4 or 8 GiB of RAM. The pi 4 model B initially came with a 1GiB model which became available on the launching day in June 2019 but the 1GiB model was discontinued in March 2020, and they later introduced the 8 GiB model in the month of May 2020.

Operating systems used by the Raspberry pi

The Raspberry Company devised an operating system for the Raspberry pi called the **Raspberry pi OS**, referred previously as **Raspbian** and a **32-bit Linux OS** primarily for download. There are also some third parties OS such as **Ubuntu, Windows 10 IoT**

Core, RISC OS and **OPENELEC.** The Raspberry pi is at the core of **Python** and **Scratch** which serve as the major programming language. The various operating systems being used by the Raspberry pi 4 are usually installed on an SD card or a micro SD. The SD card is usually located at the bottom of the board.

Installing Raspbian on Raspberry pi 4 using NOOBS

Raspbian is the most common operating system, which is installable on all Raspberry pi models. The Raspbian operating system provides your Raspberry pi the feeling of a personal computer including programming interface, wide array of browsers and many other computer programs. To install Raspbian on the Raspberry pi, you need an application called NOOBS (New Out of the Box Software). The NOOBS is an OS installation manager that makes it especially very easy to install Raspbian Operating System. There are some Micro SD cards that come loaded with NOOBS and you only have to fit it inside the Raspberry to begin. NOOBS allows Raspberry beginners to choose their preferred operating system and then install it on the Raspberry pi. You will need your Raspberry pi 4, a computer that has an SD card slot and an SD card of about 8GB to get started. Follow these guides to install NOOBS on the Raspberry pi;

- **Downloading and extracting NOOBS:** A computer is required to copy NOOBS onto an SD card. Therefore, the first thing you must do is to get NOOBS on your computer

- Navigate with your preferred web browser to the NOOBS download page at https://www.raspberrypi.org/downloads/noobs/.
- From the NOOBS download page, you will be prompted to select between downloading NOOBS and NOOBS lite. Note that you can choose any of the two, but it is advisable to select the NOOBS option because you will be able to install the OS without needing internet access. While an internet access is required to install the NOOBS lite.
- Tap on download NOOBS as a zip file.

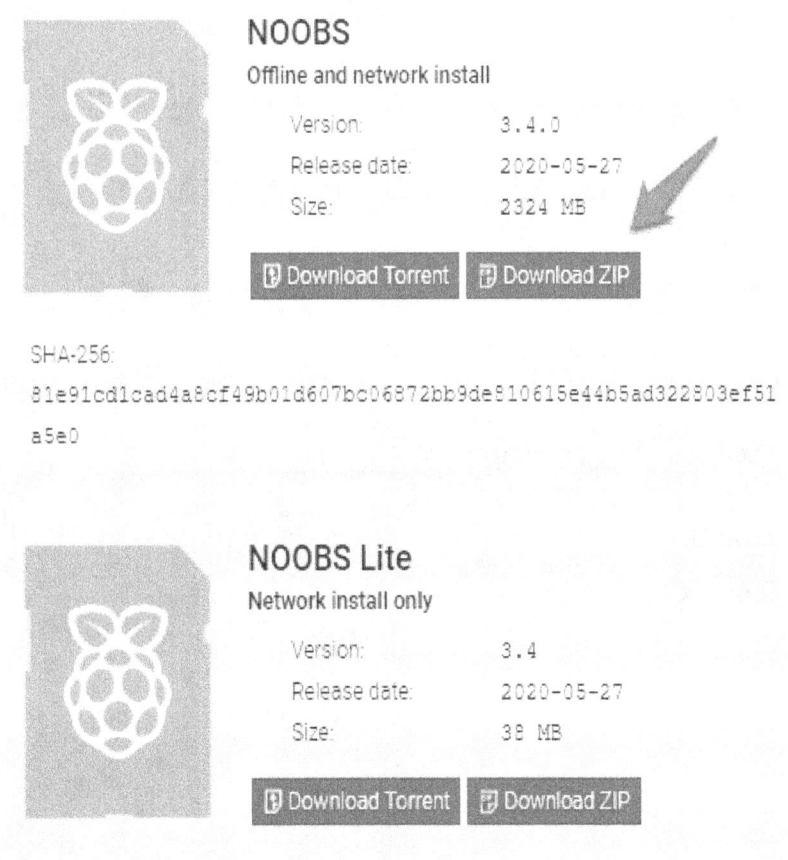

- o The NOOBS file will then be downloaded as a zip file on your computer. Go to the location where the NOOBS file has been stored on your computer and extract the file.
- Put your SD card into the right slot inside your computer. But you will need to format the SD card once it gets inside the computer. if you don't have an SD card formatting tool on your computer, download and install an SD card formatting tool (SD memory card formatter) from https://www.sdcard.org/down-loads/formatter/eula_windows/. Let the "Format size adjustment" option be set to "on." Then erase it in FAT (or MS-DOS) format.

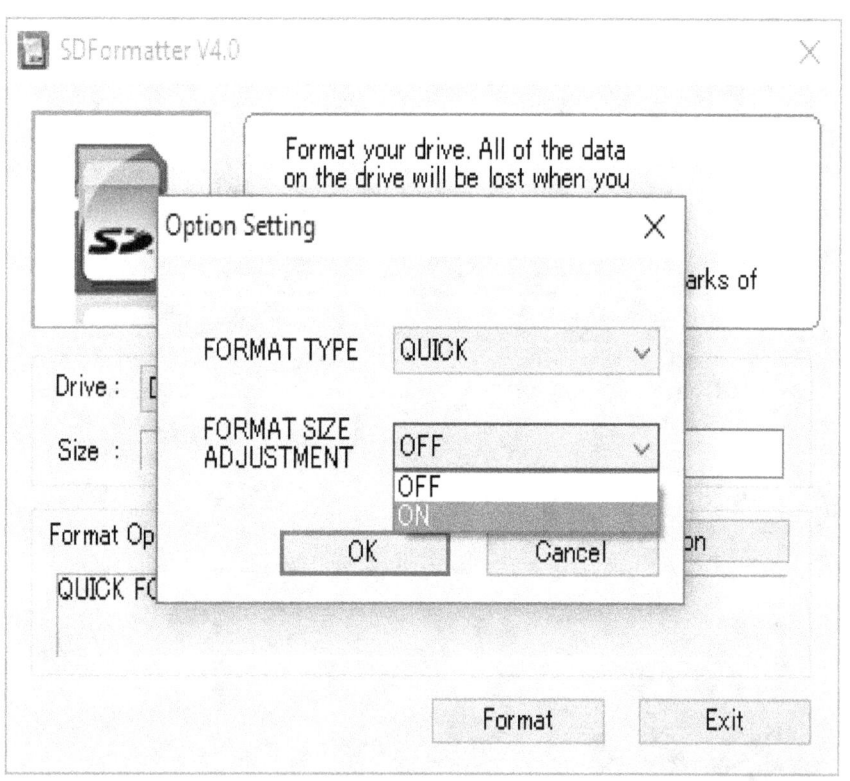

- Copy the extracted NOOBS file inside the SD card you have formatted. If the NOOBS file has been saved inside a folder, just open the folder and copy the extracted NOOBS file from the folder then paste it inside the formatted SD card.
- Insert the SD (which contains the NOOBS file) into the Raspberry pi 4.
- Once the Raspberry pi 4 boots to NOOBS, a welcome page will be prompted asking you to choose the specific operating system that you will like to install on the Raspberry pi 4.
- Select **Raspbian** from the list of operating systems displayed and select **install** to start the installation process.
- After successful installation, your Raspberry pi 4 will boot to the Raspberry pi 4 operating system.

Installing Raspberry on the Raspberry pi 4 by writing a disc image

- Navigate your favorite web browser on your computer to download the Raspbian disc image from the Raspbian website at https://www.raspberrypi.org/ downloads/ raspberry-pi-os/.
- From the website, choose the "download as zip" option to start downloading the Raspbin disc image.

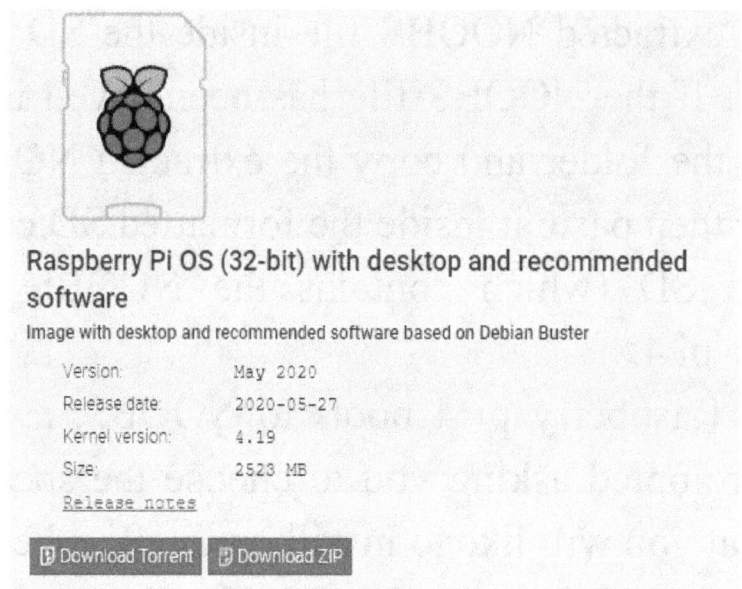

- After the Zip file has been downloaded successfully, go to the location where the file has been installed on your computer to extract the file. For Window users, you might want to use the **7-Zip** software to unzip the Raspbian file. You can get the **7-Zip** software at https://www.7-zip.org/. If you are using Mac, use **The Unarchiver** to unzip the Raspbian file. You can get **The Unarchiver** at https://theunarchiver.com/.
- After successfully extracting the file, simply write the Raspbian disc image to your micro SD card which has already been inserted into the computer. To write the disc image to your micro SD card, you need to get the **Win32 Disk Imager.** Go to https://sourceforge.net /projects/win32diskimager/ to get the **Win32 Disk Imager.**
- Once the Disk Imager has been installed on your computer, launch the software. The Disk Imager will prompt you to select the destination for the **unzipped Raspbian file** you are writing

into the SD card. Simply select the micro SD card as a destination. Tap on the double check and then click the button to rewrite the disk image.
- Insert the micro SD card (which contains the Raspbian disc image file) into your Raspberry pi 4.
- The **Raspbian OS** will now boot to the Raspberry desktop. The default username is **pi** and the password is **raspberry.**

The amounts of time your Raspberry OS will take to complete installation depend on the speed of the Micro SD card you are using. As the operating system is installed, progress is shown in a bar along the bottom of the window; you'll see a slide show highlighting some of its key features. It's important that the installation isn't interrupted as this has a high likelihood of damaging the software, a process known as data corruption. Do not remove the micro SD card or unplug the power cable while the operating system is being installed; if something does happen to interrupt the installation, unplug the Pi from its power supply, then hold down the **SHIFT** key on the keyboard as you connect the Raspberry Pi backup to its power supply to bring the NOOBS menu back up. This is known as recovery mode, and is a great way to restore a Pi whose software has been corrupted to working order again. It also allows you to enter the NOOBS menu after a successful installation, to reinstall the operating system, or install one of the other operating systems.

Getting Familiar with the Raspberry Pi 4

Configuring the Raspberry pi 4

After successful installation of the Raspbian OS into the Raspberry pi 4, you will get a welcome window welcoming you to the Raspberry desktop

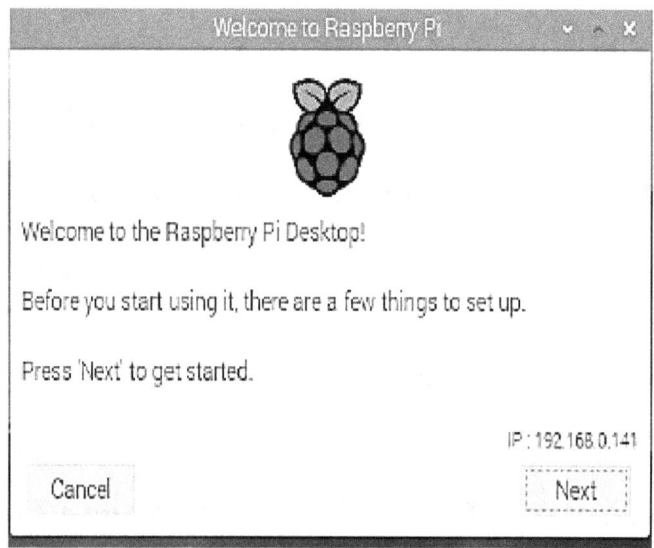

You can decide to close the welcome page above by tapping on the **cancel** tab. But you need to address some basic information so that features of the Raspberry such as wireless network will work. When you tap the **next** tab in the image above, you will be taken to a window where you can select your country, time zone and language. You will also see a dropdown where you will have access to customize the keyboard layout you want Raspbian to be using. You can customize the Raspbian desktops and programs to start showing in English, tap on the **"use English language"** checkbox to allow you set English language as your preferred language and then tap **Next** to finish.

Remember that your default Raspberry password is "raspberry" which is not secured enough, so a screen will be prompted asking you to change this password. Endeavor to input a more secure password that you will be able to remember. Tap the **next** button when you are okay with the password.

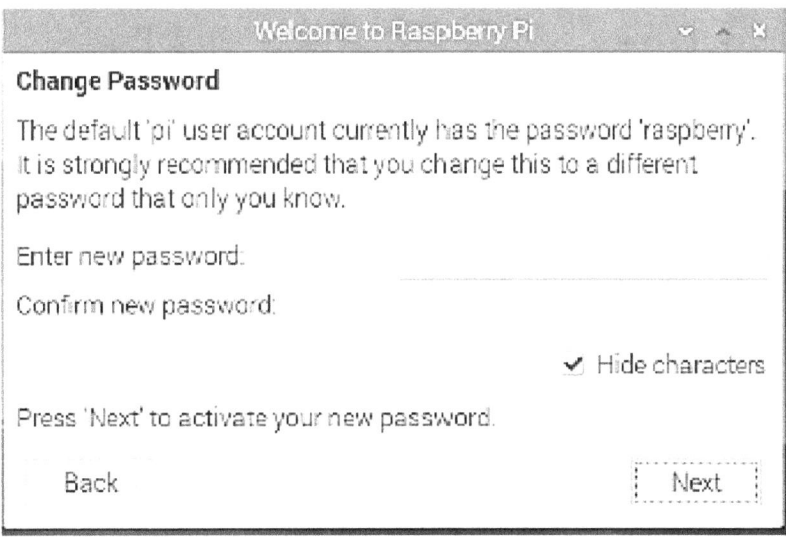

Another screen will be prompted that will enable you to choose your preferred Wi-Fi network from a list. Check the list of networks, find your network's name and tap on it, then choose

Next. If your chosen wireless network is secure (which is actually expected), you will be prompted to enter its password, which is its pre-shared key; check for the pre-shared key which is usually inscribed on a card with the router or on the bottom of the router. Tap **Next** to connect to the network. If you prefer not to connect to a wireless network, simply tap **Skip**. Note that Built-in wireless networking is only available on the Raspberry pi 3, pi 4, and pi Zero W families. If you want to use another model of Raspberry Pi with a wireless network, you will need a USB Wi-Fi adapter.

Another screen that will be displayed will enable you to check for updates and then install those updates for Raspbian OS. The Raspbian OS is designed to update itself on a regular basis which enables it to fix bugs, modify existing updates to add new features, and also improve software performance. Tap Next to install these updates, otherwise, click Skip. You will need to be patient while downloading the updates as the downloading process might take a little time to be completed. When the updates have been completely installed, a window will pop up telling you that 'System is up to date'; tap the OK button to proceed.

The final screen will be prompted where you will be able to reboot the Raspberry pi. Any change you made will only be effective when you restart the Raspberry Pi, a process known as rebooting. If you are asked to reboot, select the Reboot button and your Raspberry Pi will restart.

Exploring the Raspbian desktop

The graphical interface of the Raspbian installed on the Raspberry pi is called "Raspbian with Raspberry pi Desktop." The entire Raspbian desktop is taken up by the **wallpaper** which will serve as the background for all the programs and software run on the Raspberry pi. The **taskbar** is located at the top section of the desktop which enables you to load any program you want to run. The program you are currently running (which is usually shown on the wallpaper) will be displayed at the top left section of the taskbar. Let us say you are deleting an item; the action will be displayed over the desktop background (wallpaper) and you will see the "deleting an item at the top left section of the taskbar).

The extreme right hand section of the menu bar contains the **system tray** which might contain the following; **a media eject symbol** (you will see this symbol only if you insert a removable storage inside the Raspberry pi. If you want to remove the storage, simply tap on this symbol for safe removal), a **clock** situated at the far right hand section of the system tray (you can tap on the clock symbol to access the calendar), the **speaker icon** (you will be able to either increase or reduce the Raspberry pi volume when you left click on this icon with your mouse), a **Network icon** (The signal strength will be shown as series of bars when you are connected to the Raspberry pi using a wireless connection. But if you are connected with the pi with a wired connection, you will be able to see two arrows; one pointing up and the other arrow pointing downward. When you tap on the network symbol, you will be able to access the list of wireless networks within your location) and a **Bluetooth icon** which will enable you to connect to any Bluetooth device that is within range).

The left navigation pane of the menu bar contains the **Launcher** which contains all the programs that have been installed alongside with Raspbian (remember that Raspbian is an operating system and some programs come by default when you installed the OS just like you will have some programs installed automatically with your Windows 10). You will be able to access most of these installed programs as a shortcut while a handful of them are hidden and you can bring them to view by tapping on the Raspberry symbol located at the far left section of the menu bar.

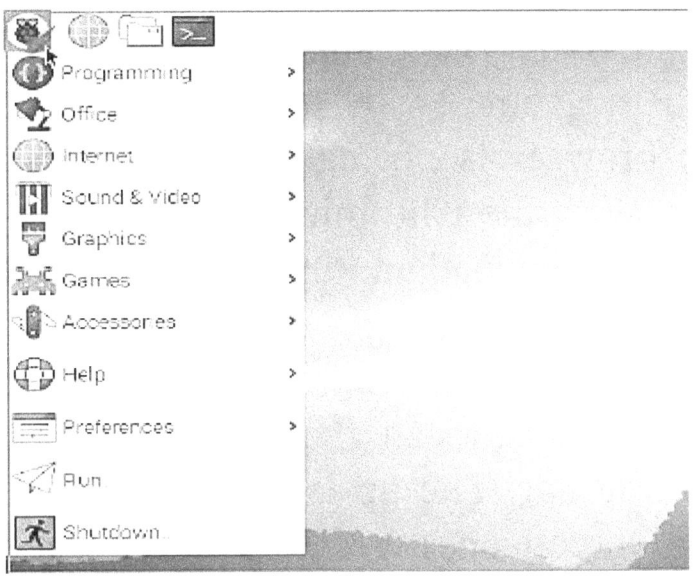

When you tap on the Raspberry icon at the extreme left of the menu bar, you will be prompted with a list of categories containing some programs. For instance, when you select the **programming category,** you will be able to access some software which will help you to write your own program as you desire while the **internet category** contains some list of programs that will enable you to access the internet.

Using your Raspberry Pi 4

Browsing the internet with the Raspberry pi 4: The chromium web browser

The chromium web browser on the pi 4 has the interface much like the regular chrome browser on your desktop computers or PCs. You can start the chromium web browser by tapping on the Raspberry icon at the top left section of the menu bar to access the internet category. Tap the internet category and select **Chromium Web Browser** to begin. When you launch the chromium web browser, you will get a welcome with "welcome to chromium." The Chromium web browser is not exactly different from the normal chrome browser. With the Chromium browser, you will be able to access the internet, download and install other software, play games and read from the media. To get more from the Chromium web browser, try and maximize the browser so that it can take up the full desktop space. To do this, locate the three icons at the top section of your chromium title bar. The up-arrow symbol (which is found at the middle of the three icons) will help you to maximize the screen to take up the whole desktop space. The minimize button (down-arrow icon) is found to the left of the up-arrow icon (maximize button) and will enable you to hide the chromium web browser's screen to top of the task bar where you can always restore it when needed. The cross icon (at the left side of the maximize button) will close the chromium web browser's window if you tap it. You are not advised to close the chromium web browser interface when you still have some important tasks that you have not saved. When you first load the chromium web browser, you might see some tabs located along the top of the chromium window. You will be able to navigate to a different tab by tapping on the tab you want to switch to. You can also close any tab by tapping the cross symbol located at the right edge of the tab; this will not close the whole chromium window. You can open a new tab, which you can use to access many websites on the chromium web browser without necessarily opening multiple chromium web apps, simply tap on the tab button located at the right of the last tab in the list, or alternatively hold down the **CTRL** key on your keyboard and press the letter **T** key before releasing the **CTRL key**. When you are done navigating the Chromium web browser, tap the close button at the top-right of the chromium window.

Saving your files on the Raspberry pi: The File Manager

Much like your everyday desktop computer, the Raspberry pi 4 can save your files so that you can access them at a later time. You can store your videos, audios, images or any other files you downloaded from the chromium web browser inside the pi's file manager. Every file you saved is automatically stored in the **home directory.** To access the home directory, select the Raspberry icon located at the extreme left of the task bar. Scroll down and select **Accessories** and then tap on **File manager.** The **file manager** will enable you to access all your folders and their files located inside the pi' SD card. You can also access files on a flash drive when you connect it to the Raspberry pi. These files will automatically enter into the home directory which houses a number of other sub-directories. These sub-directories include; **Desktop, docu-ments, downloads, MagPi, Music, Pictures, public and videos.** For instance, your music files are stored inside the **Music** subdirectory while the images are stored in the **Pictures** sub-directory. The File Manager section is divided into two separate panes: the left pane displays the directories on the Raspberry Pi, while the right navigation pane displays the files and subdirectories of the directory you selected at the left pane. If you insert a removable storage device, such as a flash drive, inside the Raspberry Pi's USB port, a window will show up requesting if you would like to open the file in the File Manager; tap OK and you will be able to access the device's files and directories.

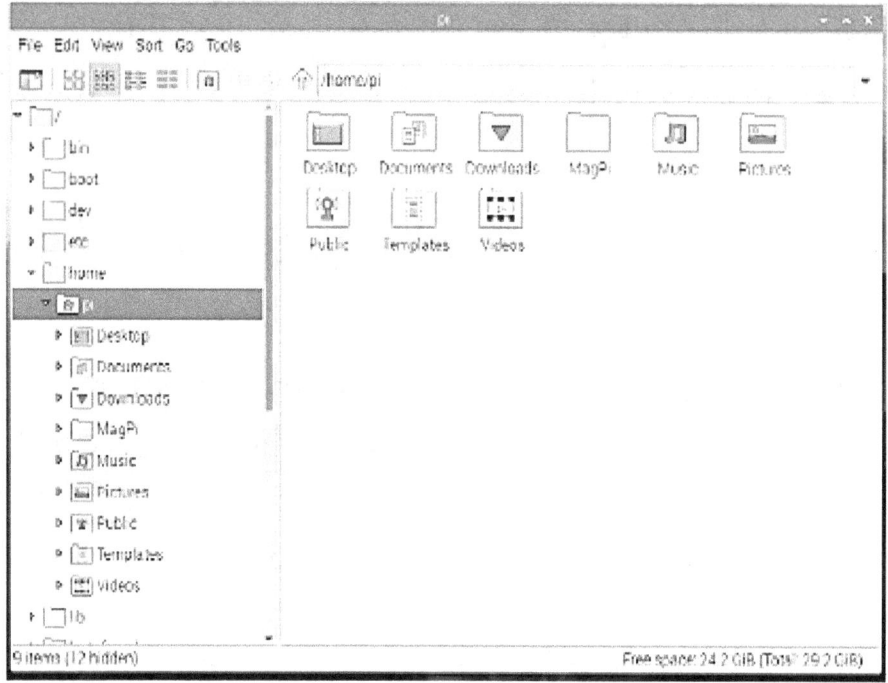

You can conveniently copy your files from a removable disk, such as a flash drive, into the pi's SD card. You can also copy files from the SD card into a removable device (flash drive). To get started, open your home directory (on the pi) and the removable item (flash drive) separately. From your pi's directory, navigate to the particular files or folders that you want to copy and then right click on the files or folders and tap **copy** or press **CTRL C** (if there are many files or folders that you want to copy selectively, simply tap on one of the folders and then scroll down while holding the **CTRL** key to select and highlight other folders or files). Open the second window, which is the removable device's window, and paste the copied file by right-clicking and then choose copy. If you select the **move** option in the SD card, this will delete the file from the SD card and the file will be available anywhere you move it to on the removable device. You can alternatively use the drag and drop method. To use the drag and drop method, tap on the file or folder that you want to drag to another location inside the removable item, and then open the removable item's window to drop the file or folder by taking your hand off the mouse. The shortcut for copy is **CTRL + C.** You can paste a file by tapping the **CTRL** button + **V.** When you tap **CTRL + X,** you are about to cut the file you have highlighted. To get more from the **File manager** window, try and maximize the platform so that it can take up the full desktop space. To do this, locate the three icons at the top section of your file bar. The up-arrow symbol (which is found at the middle of the three icons) will help you to maximize the screen to take up the whole desktop space. The minimize button (down-arrow icon) is found to the left of the up-arrow icon (maximize button) and will enable you to hide the file manager screen to the top of the task bar where you can always restore it when needed. The cross icon (at the left side of the maximize button) will close the file manager's window if you tap it. You are not advised to close the chromium file manger's interface when you still have some important files that you have not saved or copied. When you are done navigating through the file manager, you can tap the close symbol located at the top-left of the file manager's window. If you have more than one file manager window open, close them one by one. If you attached a removable storage device to your Pi, eject the device by tapping the eject icon at the top right section of the screen, find the removable item on the list, and then select it before unplugging the device from the Raspberry pi.

Writing on the Raspberry Pi 4: The LibreOffice Productivity Suite

The LibreOffice writer does the same job that the Microsoft office suite (MS word) on your desktop computer will do. To have access to the LibreOffice Suite, navigate to the pi's menu at the left section and then select Office and from the Office dropdown, tap on the LibreOffice writer. If the LibreOffice writer has not been installed automatically with the Raspbian operating system, you can install it on your own by running a simple command *"**sudo apt install libreoffice.**"* You will need a disc space of about 649MB to fully install and enjoy the LibreOffice writer. The writer is the word processor that comes with the LibreOffice. The writer is very identical to the MS word and it is used essentially for creating and editing word documents. It is easy to open the LibreOffice software with the command below;

```
libreoffice --writer [FILENAME]
```

If you specify a specific file name that you want to open in the [FILENAME], you will be able to open that file.

With the LibreOffice writer, you can carry out most tasks you do with your MS word app. You can format your document by changing the font size, font style, font color and add a powerful effect to give your texts some nice appeals. While writing with the LibreOffice writer, you can be sure of grammar correction because every misspell word is highlighted in red so that you can correct it. You can also save any file right from the LibreOffice writer by pressing **CTRL + S.**

The LibreOffice suite also features LibreOffice Calc which is just like Microsoft's spreadsheet software. The Calc software can be used to analyze both large and small data samples just like Microsoft Excel. It is easy to open the LibreOffice Calc software with the command below;

```
libreoffice --calc [FILENAME]
```

If you specify a specific file name that you want to open in the [FILENAME], you will be able to open that file.

There is another software that comes with the LibreOffice suite is the LibreOffice Impress which is just like the Microsoft PowerPoint app. The LibreOffice Impress is used specifically to create a PowerPoint presentation. It is easy to open the LibreOffice Calc software with the command below;

```
libreoffice --impress [FILENAME]
```

If you specify a specific file name that you want to open in the [FILENAME], you will be able to open that file.

The LibreOffice draw is a LibreOffiec software for making and editing pictures and diagrams.

The Recommended Software tools

It is understandable that the Raspbian OS comes with a lot of installed software, but you can still install much other compatible software. A list of some of the best software that you can install in your Raspberry pi can be accessed in the Recommended Software tool. The Recommended Software tool requires a strong connection to the internet for full access. After successfully connecting your pi to the internet, simply tap on the Raspberry menu icon at the left section of the toolbar, and select **Preferences.** Tap on **Recommended Software** to get started**.** The Recommended tool will be prompted and you can start downloading information about software that is available. After some time, the list of compatible and available software will show up. The tool will load, and you can then begin downloading information about available software. After a few seconds, a list of compatible software packages will appear. These tools, just like the software in the raspberry menu, are systematically arranged into various categories. Tap on a category in the pane on the left to access software from that category, or choose All Programs to see all the software. In the "**All programs"** section, if you see software that has a tick beside it, it shows that the app has already been installed. If there is no tick beside the software, you can tick it for installation. You can as well uninstall any software that has already been installed by unmarking the tick beside it. After selecting the list of software that you either want to install or uninstall, proceed by tapping on the **"OK"** button to start installing or uninstalling as the case may be.

The Raspberry pi configuration tools

The Raspberry pi configuration tools enable you to change or modify settings in the Raspbian OS. To access the Raspberry configuration tools, simply tap on the Raspberry menu icon at the left section of the toolbar, and select **Preferences.** Tap on Raspberry pi configuration to get started. The Software tool has four tabs which carry out different functions and settings. These include;

- **System tools:** This enables you to set the host name for your pi, change your account password and other settings. You don't really need to change many things here.
- **Interface tab:** You can add new hardware, like the Raspberry pi camera module, by changing some settings on this tab. You can also change some of the settings in this tab if the manufacturer advised it. You can set your SSH, which gives a 'Secure Shell' and enables you log into your Raspberry Pi from another computer on your network using an SSH client. You can as well set the VNC, which gives a 'Virtual Network Computer' and allows you to see and control your Raspbian desktop from a different computer on your network using a VNC client; and the Remote GPIO, which allows you to use the Pi's GPIO pins from another computer on your network.
- **Performance tab:** You can boost the performance of the Raspberry pi and also determine the amount the pi's Graphic processing unit will consume.
- **Localization tab:** Change the locale which determines how language is set in the Raspberry pi and how numbers are being

displayed. You can as well change your time zone and the layout of your keyboard in this tab.

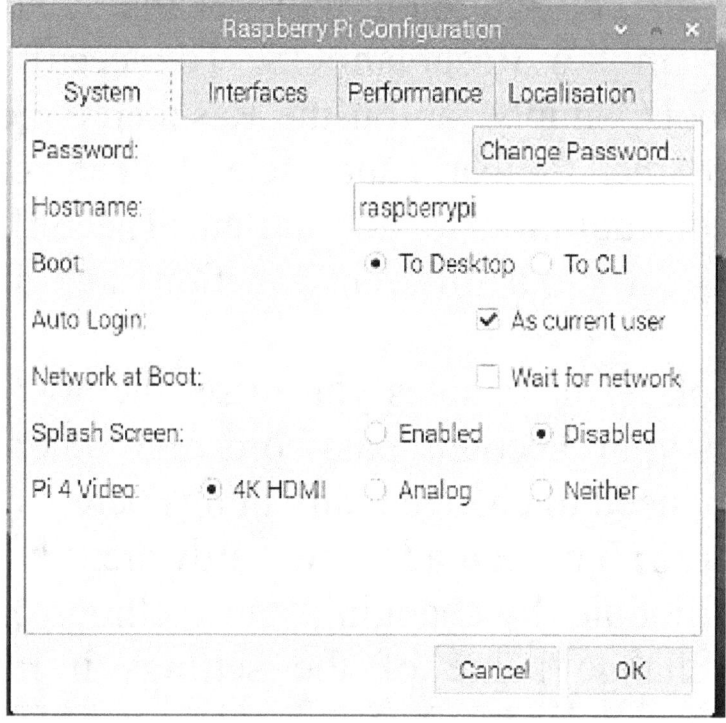

Shutting down

You can shut down the pi by tapping on the Raspberry symbol at the top left side of the menu bar. The shutdown window will be prompted allowing you to choose between shutdowns, logout and reboot. The shutdown option will close all the software you have opened and shutdown the Raspberry pi. It is recommended that you save all the files you are working on before you shutdown the operating system.

CHAPTER TWO

UNDERSTANDING THE RASPBERRY COMMAND LINE

The command line is a simple way of passing information from users to the computers. It provides a unique way of breaking down complex tasks and making even the simple ones simpler. The Raspbian operating system on the Raspberry pi 4 anchors the various functions, including the command interface on the little computer. With the command line, you don't need to press any icon or navigate extended menus on your pi. All that you basically need to do is to understand the task you want to carry out with its command. You can access the command line using a program referred to as a terminal emulator or Shell or Bash. The command lines may be a little bit stressful to learn, but once learned they give you a quick way of accessing and navigating your files and programs. The command line is called Terminal and you can find it under Accessories in the menu. Once you enter the terminal, you will be taken to the interface where you can start entering your commands for onward execution.

Command for listing and accessing files

You will be able to access and navigate files wherever they are on the Raspberry pi by specifying their paths. This is done by issuing a simple command to help locate those files. The path (folder hierarchy) is conquered by ls / – which is specifying the path. You can enter pwd (present working directory) to know your current location. Don't forget the ls / – whenever you are entering such command. The home directory for any user that is already logged in will be specified as ls ~. Note that anytime you enter a command, you need to press the **return** or **enter key** to let the bash understand that you have issued a command.

To create a directory, you will need to first confirm that you are in your home directory. Simply enter **pwd** to check if you are in the home directory. To create a new directory, type mkdir tempfolder and have a look with ls.

Installing software with the command line

You can deploy the apt command to update your pi's list of installable software. Begin by issuing the command like: apt-get update. When you issue this command as a pi user, you will be greeted with an error message. This is because the process of changing system software on a GNU/Linux (or any type of UNIX) operator is a job restricted to user with administrative permissions: like the admin, also known as root. You can bypass this restriction by using the sudo command. The Sudo command will actually offer you a perfect choice of permissions to have access to portions of the admin user's powers. However, inside the Raspberry Pi, the Raspbian OS assumes that the default user is someone that just wants to get some things done, and sudo in front of a command will verily allow you to do much things. Entering the two commands below will allow you to update the software installed on Raspbian;

```
sudo apt-get update
sudo apt-get upgrade
```

You can exercise patience for one command to end before prompting the second one or you can run both at a time by;

```
sudo apt-get update && sudo apt-get upgrade
```

The && in the command above is a Boolean (logical) AND, which implies that if the first command fails to run well, the second command will not be executed. The reason is because for a logical AND to be true, both of its conditions must be essentially true. It is always worth running the update command before installing new software, too – minor updates are made even in stable distributions such as Raspbian, to address any issues. For instance, to install a game called bsd, just prompt the command below;

```
  sudo apt-get install bsdgames
```

You can search a list of apps or games by entering a simple command; apt-cache search games. You can as well examine a game or an app with the command; apt-cache show bsdgames.

Any software, game or app that you can download are usually stored in **/var/cache/apt.** whenever you are low on disc space, you can issue a command to help you relieve the disc space. Just enter **sudo apt-get clean** to delete the archive. This command will not delete your installed software.

The file attributes are three sets of three different characters (rwx) informing you which specific user may read, write or execute the file or directory for, respectively, the user who owns the file, the group owner, and everyone else ('others'). Permissions for execution are required to run a file if it is a program – such as **launcher.sh** which initiates the Python games in your **usr/share/python_games** folder, and thus it has the x – and for directories, so that you may **cd** into them. Change directories (**cd**) into **usr/share/python_games** and then enter the command **sudo chmod a-x launcher.sh** – where **a** stands for all (user, group and others). You can deploy **u**, **g**, or **o** to switch between users, groups and others. When you try to open the Python Games from the main menu after all these modifications, it will not work. You can restore back to normal functioning by entering **sudo chmod a+x launcher.sh**, but instead you should deploy: **sudo chmod 755 launcher.sh**. You can deploy the command **chown** to change the owner of a file and use **chgrp** to change which group the file belongs to. If you initiate a new text file and prompt **sudo chown root myfile.txt** – and you then try to edit you will observe that while you can still read the file, you will no longer be able to write to it. You can also make a file that you can write to and run, but you can't read it any longer.

Personalizing the command lines

The Raspberry terminal contains the interface pi@-raspberrypi ~ $. The $ signifies the dollar prompt awaiting you to enter any command. The ~ is the notation for home specified by **/home/pi.** The @raspberry denotes the username which is the computer name in this case. All these signify your identity as a user and you can either personalize or change them whenever you want.

- **Creating a new user account**

If you have family members or friends sharing your Raspberry pi with you, you need to change the username from **pi** and assign a name for each individual member. You can easily add a new user account with the command Sudo adduser username. Let us say you want to create a user account for your child named Rose; you can just prompt Sudo adduser rose. You will be asked to enter a password – make sure you enter a strong password that you can easily remember. You can have a look at **/home/rose** and you will notice the home has no content since the username is new to Raspberry. To change the password for **Rose,** simply prompt Sudo passwd rose. If you are logged in to pi as the owner, you can simply prompt passwd and you will get a window asking you to change your password. You can switch your identity as pi owner to any other user, let say **Rose,** just prompt Su – rose. This prompt will take you to **/home/rose.**

Command lines for managing removable disc

When you slot any removable disc inside the Raspberry pi, the Raspbian operating system will ordinarily prompt you to open the item. But you might need some command lines to get more out of the Raspbian-removable disc interface. It is easy to plug any removable disc inside the Raspberry pi, but that is not enough as you need to make the disc available (mounting the disc) so that the Raspbian OS will be able to read and understand the content of the files inside the disc. Enter **mount** to access some of the files on the disc. The Pi SD card (partitioned as **/dev/mmcblk0p1**) is usually separated and partitioned from the removable disc (partitioned as **/dev/sda1**). You can un-mount the SD card with a **Sudo** command. Just enter the prompt **sudo umount /dev/sda1** (note the umount is without "n")

Command line for network

If you want the pi to boot with the same IP address each time, follow the steps below to edit the file **/etc/dhcpcd.conf**.
- Enter the command **sudo nano /etc/dhcpcd.conf** inside the command prompt. Go to the bottom of the script and insert the following lines;
  interface eth0
  static ip_address=192.168.0.2/24
  static routers=192.168.0.1

```
    static domain_name_servers=192.168.0.1 8.8.8.8
    interface wlan0
    static ip_address=192.168.0.2/24
    static routers=192.168.0.1
    static domain_name_servers=192.168.0.1 8.8.8.8
```

Save the file by pressing **CTRL+O** and you can then exit nano by tapping **CTRL+X**. the Raspberry Pi will now start booting up with the IP address 192.168.0.2 every time you initiate booting; we are not using the IP 192.168.0.1 as this IP is reserved specifically for the router. You can actually use any address you deem fit, but in the above configuration, the range must be between 192.168.0.2 and 192.168.0.255. Reboot with sudo shutdown -r now. Log in and type hostname -I. You should be able to see the IP address you set in the etho or wlano entry (192.168.0.2).

Command lines to stop an ongoing process

You do not need to turn off the Raspberry pi just to stop a process you have already initiated. You can run some simple commands to truncate any process you initiated. You need to identify the process first by entering the ps command. Enter ps auxww to check – the a and x insertion give all processes, the u checks for processes by a particular user, the w gives wider output. The ps aux listing features many headers, including the USER who is the owner of the process, and the PID (Process Identification number). The process identification number begins with 1 for init. It becomes especially very easy to stop a process when you know the process's PID, if that is the only way of shutting it down. For instance, to interrupt a program with a PID of 4098, simply enter **kill 4098**. You also prompt **killall** to kill a specific program by name. For instance, you can kill the chromium browser by **killall chromium browser.**

Command line to put a program in background

You can put any program or app in the background by simply inserting an ampersand symbol (&) at the back of a command in the shell. If you decided to put, let say, an app called bds in the background, simply try with **bds &** and you will receive an output like: **[1] 12768**. The first number indicates a job number, which was assigned by the shell, and the second number is the Process Identification number which we have already talked about. The bds app will now be running in the background, and you will be able to use the job control number to control the process in the shell. Initiate some other processes in the background if you want (by appending **&**), then bring the first – bds app – to the foreground with **fg 1**. Now you should see the bds app running again. You can as well put a running shell program in the background by 'suspending' the program with **CTRL+Z**. **fg** will usually bring back the most recently suspended, unless you indicate a job number. Note that these job numbers apply only within the shell where the process started. Type **jobs** to see background processes; **jobs -l** adds in process IDs (PID) to the listing.

Accessing the Raspberry pi remotely from other PCs using SSH

The SSH can allow you to use your Raspberry pi remotely at any time. To start with SH, endeavor to change your login password to **passwd.** The SSH server is not usually accessible on some Raspberry pi, you can enable SSH by prompting the command sudo raspi-config, then navigate to ***advance settings*** and choose to enable SSH. You can check the IP address assigned to the Pi using ifconfig. This will allow you to connect to your Raspberry remotely from another computer using your SSH. If you want to connect to your Raspberry pi from a Mac computer or GNU/Linux computer, deploy ssh from a terminal to allow you to connect to the Pi. Assuming a default setup, and ifconfig revealing an IP address of 192.168.0.2, connect with ssh pi@192.168.0.2 and then input your password. You can also use the OpenSSH client on Windows 10 computers; for earlier PCs, install an SSH client like PuTTY **(magpi.cc/uLytfk).** Android users will be able to establish connections using the ConnectBot client.

CHAPTER THREE

THE RASPBERRY PROJECTS: BUILDING WITH THE Pi

The Raspberry pi community contains groups of like minded individuals who are innovators, developers, engineers and internet experts who are building great and amazing things with the Raspberry pi. This chapter will explore some of the amazing projects that have been done with the Raspberry pi.

The Mathematica Telescope

The Mathematica telescope is one of the numerous projects developed with the Raspberry pi. The Raspberry pi can be used to access a wide array of mathematical data and knowledge. A very popular astronomer from Wolfram Research institute, named Tom Sherlock, used the Raspberry pi as a telescope to explore the sky right from his backyard. In this project, Tom attached the Raspberry to the eyepiece of the telescope. This is because the Raspberry pi has a camera module that can be used to take good pictures. The Raspberry pi, on its own, cannot take extended photographs of the moon. So, Tom improved the pi's image taking-ability by using some extra language and Mathematica. To make the telescope start seeing stars, the following procedures are involved;

- Set up the telescope by mounting it accordingly.

- Attach the telescope to the Raspberry pi. A serial-to-USB adaptor can be used for this purpose. You can then boot up the pi.
- Once the Pi is booted up and in Mathematica, you can start issuing commands using the Wolfram Language to control the telescope – this will change depending on what mount you're using.

Poopi and piter's paludarium project

This project was used to stimulate and recreate what looks like an Amazon rainforest. This project can be used to stimulate and recreate various time of the day and different weather conditions. The Poopi and piter's paludarium is controlled essentially by the Raspberry pi and ATMega 168Ps (four of them). The features in the paludarium with their functions include; 6× independent sections of halogen lights, 27× independently controlled 1W LEDs for **various effects**, 3× independent 3W RGB LEDs for **ambient color effects**, 3× independent 3W LEDs for **thunder and moon simulation**, 3× independent 10W LEDs for **paludarium lighting**, 2× independent fans for **wind simulation**, 3× fog generators, and 2× independent solenoids for **rain control Temperature monitoring.**

Digital Zoetrope

This is a pre-film animation gadget that gives the illusion of movement with still images. The Digital Zoetrope is controlled by hand. Like the original designs, you spin the device and look through the slats to see movement in the still images as they rotate.

This project uses 12 OLED displays with the Raspberry Pi, and it is actually possible to update the frames in real-time, so you could watch an entire film if you wanted. Moreover, using technological trickery, it is possible for two people to view entirely independent animations when looking into the Digital Zoetrope from different angles.

Internet of Things Chessboard

This is a real physical chessboard that could play online. It is different from online games where you can only play the game in 2D view which makes the game a little bit unintuitive. To play, your opponent requests a new game on the chessboard's own website. When they make a move by dragging a piece on the web interface, the relevant piece and square will flash on the IoT Chessboard. Whoever is sitting in front of it will move that piece to update the board, then will make their own move. The board detects this, and will then send that move back to the opponent's web UI, where the piece is moved on the screen. The Internet of things chessboard utilizes the Raspberry pi as its web server and deploys an Arduino MEGA to direct the electronic part. Each of the 3D chess pieces comes with a 3mm jack which acts as a power source for the LED located inside the chess piece and allows you to know where the pieces are located.

Naturebytes wildlife cam kit

The Naturebytes Wildlife Cam Kit uses a Raspberry Pi together with a Raspberry Pi Camera Module that is geared by a PIR motion sensor, to enable you to take pictures of birds and animals lying around in their natural habitats.

LIFEBOX

The lifebox can be used as a nightlife. In the lifebox live two pixelic entities, the blue and yellow species. These two species compete to survive and reproduce, feeding with the white mana that grows under their feet. Each species has eight configurable variables that can change their behavior. The white mana also has five parameters that determine their behaviour and also rule the future of the two species that feeds. LifeBox, in short, is a virtual ecosystem simulator on a 32×32 RGB led panel. It is composed of two species that compete for the resources (mana) to get energy, survive, reproduce, and grow. Both species and the mana (which actually acts as a different species itself) have a user-defined parameterization that allows the user to change their behavior and see the consequences on the panel, acting as a god of the virtual ecosystem. The lifebox features a Raspberry pi, a LED panel plus a driver which connects the LED panel and the Raspberry pi. Since the lifebox contains two entities - LED and a pi – the code is divided into two separate sections; a part that controls the driver and the section controls the stimulation.

JOYTONE

The Joytone is an excellent musical gadget featuring a hexagonal grid that contains 72 joysticks. The Joytone's hexagonal grid reveals musical designs and patterns that are usually hidden by the eccentricity of some common audio-style interfaces, such as the white and black keys in a piano. Each joystick plays one note at a time, and the motion of the joystick determines the capacity (volume) and character of the note. The notes are distributed over the hexagonal grid which shows that all forms of musical patterns are not ambiguous.

PiScan

The PiScan is just an open-source version of Amazon Dash. With the PiScan, it is easy to scan any product you want to buy using a Raspberry Pi and order them directly. It is hugely exhilarating and very practical, and it is in fact more futuristic than Amazon dash which is Amazon's official device. The PiScan is the ultimate solution for lazy-person shopping. What the PiScan does is to read the barcode on any consumer product you want to buy and then order the product for you from an online vendor. PiScan changes scanned items into a list that you can use to order products, using Amazon's API. When it comes to hardware, PiScan uses a Pi Model B together with a Wi-Fi dongle and a USB laser scanner that can scan barcodes. The USB scanner is the only extraneous hardware incorporated in the setup, and you can get one up from Amazon for less than £20. There is software that instructs the Raspberry pi to listen for inputs from the USB barcode scanner. The barcode scanner works just like a keyboard, except its input comes in short bursts of characters. The input is a 10- to 13-digit number matched to the Open Product Data database (product-open-data.com). If the barcode scanner sees a match, it will enter the identity of the product into a list. The Raspberry Pi delivers the list to you as an email, or you can tick items to add to your Amazon shopping cart.

- Laser USB scanner: The only piece of hardware incorporated inside the setup is the Laser USB scanner from Amazon. The Laser USB scanner is deployed to capture barcodes from products and sends the digital information to the Raspberry Pi (which then matches it to a database).
- Software installation: With the scanner connected, you can then proceed to install the software.
- Scan and shop: The product can then be scanned with the barcode. The barcode is matched with an already available database of products. The product will now be marked and you can then shop for them on the online platform.

Watch iPlayer on Raspberry pi

Requirement: SD card worth 8GB size, broadband connection, television or any other screen (monitor) and some line of codes.

Get_iPlayer is an excellent, open source utility software which enables users to access and stay updated with what is currently streamed on the BBC's iPlayer web. With the Get_iPlayer, you will get to download any TV programs of choice. You will also have the chance to choose your video according to the device you are using to view it. Some devices cannot access videos in resolution higher than 512x288, while most other viewing devices can go as high as 1280x720. The BBC's website claimed that you can only keep the files you downloaded or streamed for just a period of 30 days. But you should not worry as there is no DRM on the videos you watched and downloaded, and as such nobody is personalizing you for watching.

Step 1- Update your package lists: You are going to install some software on your pi. When you install any software which is anchored by the Raspbian OS, you need to first navigate to the command prompt. At the command section, prompt the command **sudo apt-get update** and then execute by pressing ENTER. This basic command prompt will help you to update your software list. For the second time, prompt the command **sudo apt-get update** and then execute by pressing ENTER once all the software have been updated to their latest configuration. The second time command is for validation. The next step is for you to add Jon Davies's personal package archive to your **sources.list.** Go to github.com/raspitv/get_iplayer/blob/ master/code.tx to copy the five lines of GitHub code and then paste the codes inside a terminal window on the pi.

Step 2- installing the keyring and software: You will need to repeat the **sudo apt-get update command** in the Raspberry pi terminal. It is likely you get an error message for keyring, hence you will need to install Jon's Keyring. To install the Jon's keyring, prompt the command sudo apt-get --allow-unauthenticated -y install jonhedgerows-keyring and then hit ENTER. You should repeat this sudo command the second time as you have done in the previous step. You can then proceed to install the Get_iPlayer on your pi by prompting the command sudo apt-get install get-iplayer.

Step 3- using the Get_iPlayer: Once you have successfully installed the Get_iPlayer, you can proceed to get more information about the i_Player to enable you to get the best out of it. To begin, enter the prompt get_iplayer –usage in the terminal which will deliver you some list of options to choose from. Prompt the command get_iplayer –help if you want to request for more options. You can as well utilize the prompt get_iplayer –longhelp to get further options. Before you start downloading any program, you should collect the index of all the available contents. You can do this by prompting the command get_iplayer in the terminal.

Step 4- filtering your downloads: Sometimes, there are just too many things to download and you might get a little bit confused. Simply filter the list by prompting the get_iplayer command in the terminal and then use categories together with the command. You can select any category you want from the list; drama, documentaries, Science &technology, sports etc. You can prompt get_iplayer --drama to get a drama list.

Step 5- downloading your items: When you take a look at the category of films displayed, each film begins with a program ID number (PID). Let us say you want to get the film **breaking bad** with a PID 3467 and you intend to download the film using the best available resolution, say 1280x720. You will just need to prompt get_iplayer --get 3467 –modes. The file will be downloaded after a few minutes and it will be processed into an MP4 which can be accessed, saved and deleted.

Step 6- watching the items you have downloaded: There is a GPU-accelerated media player which comes by default when you installed the Raspbian operating system on your Raspberry pi 4. This media player is called **omxplayer** and it can access videos in HD even in the latter generations of Raspberry. To watch the item you have downloaded, simply prompt the command omxplayer [filenamehere].mp4. If the file name is long, type the first few letters of the filename and tap the TAB key so that the file will be completed automatically. Then press ENTER.

Shooting in Slo-Mo with the Raspberry pi camera module

Requirements: pi camera module, fast internet connection and an event to shoot.

The Raspberry pi camera module can be used to shoot videos in slow motion configuration and then convert the video into format that is accessible on all devices. Slow-motion video is not a new concept as cinematographers have been shooting videos in slow motion for a long time. When you see an actor in a movie jumping off from a three-storey building, the slow motion is in play because the actor is exactly jumping that fast in reality. It is just like a walk in the park to create your own slow motion video using the pi camera module. What is involved is just some short code listing that will enable you to shoot your video and then convert it into a desirable format so that it can be accessible on all devices.

Step 1- Connecting the pi camera module: This is about the very first thing you need to do. Get started by first turning off your Raspberry pi. Be especially careful with the camera module while handling as it is actually sensitive to static – ground yourself by touching something (such as radiator) before you get started with the camera module. The cable of the camera module will be connected to the socket in the Raspberry pi at a location very near to the HDMI port. Ensure that the metallic end of the pi camera cable faces the metal HDMI port.

Step 2- camera module configuration: For Raspbian wheezy user, prompt the command sudo raspi-config in the pi terminal to input the configuration menu. With the arrow key, navigate down the list that shows up and choose 'Enable Camera' using the right arrow key. A new menu will be prompted where you can then choose 'Enable' with the right arrow key to turn the camera module on, and then tap RETURN. If you are using Raspbian Jessie, navigate to Preferences and choose Raspberry Pi Configuration from the prompted menu, then enable the camera from the Interfaces tab. Reboot the pi if you are asked to do so or you can alternatively enter the prompt sudo reboot in the terminal to start the rebooting process.

Step 3- install a video converter: The default recording mode for the Raspberry pi is the raw H.264 files. This file system doesn't work on some devices and it is advisable to change to another configuration. To convert the H.264 file mode into a playable mode by installing an app called **gpac**. To get started, enter these commands; sudo apt-get update followed by sudo apt-get install gpac, and then follow the instructions displayed on the screen.

Step 4- testing the camera: To ensure that the camera is working perfectly, test the camera with some commands. Connect a screen to the camera and prompt raspistill -o test.jpg in the terminal window. The picture should be displayed on the screen for a little while and the picture should be saved you're your home directory if everything is working fine. If the picture is not showing, check to confirm that you entered the correct command, or alternatively turn off the pi and reestablish connection with the camera cable.

Step 5- write a python script: python is one language that can be used to create a Slo-mo video script. Open a text editor in Raspberry (you can use the Leaf text editor which has been installed automatically with the Raspbian OS) and then copy the code opposite while being extremely careful to avoid copying a misspell code. You don't have to copy the comments in the code (lines starting with #) because the Python language doesn't use them. The script deploys the OS Python library to carry out the terminal commands that you have entered. Save your file as **slowmotion.py** in your Home directory (/home/pi).

The python script is as it is below;

```python
import os

import time

print("Starting program")

time.sleep(2)

##### Record the slow motion video #####

# '-w' sets the width # '-h' sets the height

# '-fps' sets the frames per second (90 maximum - for slow motion)

# 't' sets the time in milliseconds (30000 = 30 seconds)

# '-o' sets the output filename

print("Recording started - 30 seconds")

os.system("raspivid -w 640 -h 480 -fps 90 -t 30000 -o vid.h264")

print("Recording complete. Please wait...")

time.sleep(2)
```

```python
##### Convert the raw recorded video file to playable mp4 #####
# '-add' is the name of the raw video we want to convert
# The second filename is the output mp4 file
# (we use the same name followed by '.mp4')
print("Converting video. Please wait...")
os.system("rm -f vid.mp4")
os.system("MP4Box -add vid.h264 vid.mp4")
print("Video conversion complete")
time.sleep(2)
print("Closing program")
time.sleep(2)
```

Step 6- running the script: initiate the python script by opening a terminal window. Type **cd** and select RETURN; this makes sure you are right in the home folder. Enter the prompt **sudo python slowmotion.py.** The condition of the script printed will be displayed as it is performing the command. The camera module's LED light will turn on during recording. As soon as the video has been converted successfully, the script will stop running. The video can be watched on the pi using the omxplayer by entering. To run the script, simply open a terminal window, type **cd** and hit RETURN to ensure you are in the Home folder, and then type **sudo**

python slowmotion.py. You will see the status of the script printed in your terminal window as it carries out its commands, and the Camera Module's LED will light up while it's recording. The script will end when the video has been converted. The recorded video can be accessed on your Raspberry pi by entering the command **omxplayer vid.mp4**. You can as well copy the video into many other devices or drives.

CHAPTER FOUR

PYTHON PROGRAMMING FOR RASPBERRY

Python is a free programming language, which has become the most used language being used among developers, hobbyists and students of knowledge. The python programming language works on a wide variety of platforms such as Windows (all versions of Windows), Linux based operating system and Apple OS. The python language has straightforward commands, which are very easy to understand. These commands are called syntax. These simple commands make it especially easy for beginners with zero knowledge of python to grasp the basics and move to being a pro within hours or days of learning. The python language is a high-level object oriented language with improved readability. The language is exploited by many companies such as Google, YouTube, NASA and a host of others in many of their programming related activities. The python programming language came into existence in 1991 with the help of Guido van Rossum.

Recently, the object-oriented language – python- has gone through some series of changes moving from the latter version 2 to the new version 3. The python v3 works based on UNICODE, which can operate with both English, and non-English characters, while the v2 can only work with only English characters. Python v3 is a compressed version, which makes the program smaller and very easy to understand than the v2. In this guide, you will be coming across the new version 3 of the python programming language.

Many operating systems support both the v2 and the v3. Raspbian, the operating system powering your Raspberry pi 4, also supports the v2 and v3 python programming language. The Raspbian OS comes with many graphical interfaces (which define the desktop outlook) and you can configure any of them to start using. In this python guide, the LXDE graphical interface, which is based on Linux, will be used. The Linux is the engine of the Raspbian operating system. A popular programming environment for the LXZDE is the IDLE. When you tap the LXDE program menu, you have access to a lot of features and menus. One of such menus is the programming menu (IDLE). The programming menu features the IDLE (python v2) and IDLE 3 (python v3).

Understanding the python environment

The Raspbian Linux distribution package comes loaded with python 3 and all the necessary tools you will need to make your programming easy. The following items are preloaded with v3; python interpreter, interactive python shell, a python development environment and a text editor. You can check the Python interpreter and interactive shell versions on your system by opening the LXTerminal in the LXDE Graphical UI. Type python3 -V and press Enter. If you are greeted with the message command not found, then for some reason, the Python v3 interpreter is not installed. You can also check if a Python development environment has been installed, open the LXDE graphical interface and look for the IDLE

3 icon on the desktop. If you cannot find the IDLE 3 icon on your desktop, check in the LXDE menu by clicking on the LXDE Programs Menu icon and hovering your mouse pointer over the Programming menu option. You should see the IDLE 3 icon there.

The text editor called **nano** will be required for the python programming. Check if it has been installed by opening the LXTerminal in the LXDE graphical interface, and then type nano -V and press Enter.

If any of; python interpreter, interactive python shell, a python development environment and a text editor is missing, you can install it with a simple sudo command. Just open the LXTerminal and enter the prompt sudo apt-get install python3 idle3 nano in the command prompt, and then press ENTER. This prompt will fix any of the four tools that are missen. To install them, make sure that the pi is connected to the internet and booted up.

Check the Raspberry keyboard: For most users in some part of the world, like the United States, the keyboard is not usually configured properly. If you carry out a simple check with your keyboard by pressing the **shift** + @ key and you are getting a double quote symbol instead of the @ symbol, then you need to set your keyboard so that it can work effectively for python programming by following the steps below;
- Boot up the pi, if it has not been booted on, and navigate to the LXDE graphical user interface.
- Tap the LXTerminal icon twice to access the LXTerminal window.

- Enter the prompt sudo raspi-config on the command prompt and press Enter.
- In the Raspbi-config window, scroll down to configure_keyboard and then press Enter. You need to be patient here because it may take several seconds for the next window to open.
- A new window will be prompted asking you to select the model for the keyboard, tap ENTER to continue with the default keyboard mode.
- Another window will be displayed telling you to select the layout matching the keyboard for this machine, scroll up the menu and select English, then press enter.
- On the next three screens listed, modify the selections or press Enter to accept the defaults: Key to function as AltGr screen, Compost Key screen and Use Control Alt Backspace screen
- In the Raspbi-config window, press Tab until you reach the <Finish> selection and
then hit Enter.
-Because the keyboard changes will not take effect until you reboot your system, type sudo reboot in the LXTerminal window and press Enter.
-After your Raspberry Pi reboots, test your keyboard. See if pressing the @ key now produces the symbol @ and pressing the " key produces a double quote (").

The python development environment shell is an essential tool that can be used to create, test and modify python script. The development environment shell can help you to troubleshoot any incorrectly written python syntax, and you won't need to rerun the entire python script if there is any error. The default python environment development shell in Raspbian is the IDLE.

The IDLE development environment shell

The full meaning of IDLE is interactive development environment. The IDLE features a built-in text editor and a lot of other nice features that help users to create and test python script. The IDLE 3 will be used for the purpose of the python programming. You can bring up IDLE 3 in the LXDE graphical interface by clicking on the IDLE 3 icon on the desktop twice. You can also bring up the IDLE 3 from the LXDE program menu. The following are features you can see in IDLE that will help you get started with python;

Menu-driven options and their matching control keys— You will be able to access a new IDLE window by tapping on the File menu option and then choose "New window" from the dropdown displayed. You can alternatively press **Ctrl + N** to open a new window

Basic text editor— The basic text editor allows users to enter a python script and edit it if necessary. Get started by opening a new window from the main IDLE window to have access to the text editor. From the text editor, you will be able to cut and paste any text using the control keys.

Code completion— This brings necessary hints that can guide you on how to complete the syntax you initiated. It is just like a text auto-completion

Syntax checking—When you initiate a command and press Enter, the Python interpreter checks and confirms the syntax of your statement and intimate you of any error or problem immediately. This saves time because you will be able to find out about any error during the writing stage, and the error can be corrected as much as possible.

Color coding—The IDLE shell gives color to your codes syntax as you are writing it to help you follow the logic of your Python statements.

Indentation support—Python needs the use of indentation for most of its makeup. The
IDLE shell identifies these needed indentations and handily delivers them to you.

Debugger features— You are debugging a program when you remove all of the syntax or logics that are not correct. With IDLE, the Python interpreter's syntax checking typically finds syntax errors. You can uncover logic problems by using the IDLE Debugger, which allows you to step through a program without adding additional Python statements.

Help—Because everyone needs a little help, IDLE provides a nice help facility. You can
access the help facility by selecting the Help menu option on the menu bar of the IDLE window
and clicking IDLE Help in the dropdown menu.

Note: You can leave the IDLE shell arena by taping **Ctrl + Q**. You can as well open the IDLE by pressing **Ctrl + O**.

Running a python script in IDLE

The python script comes in handy for you because it allows you to have a whole file of python statements ready, which you can run at convenience. With this, you don't need to stress yourself typing each python statement whenever you are required to run a program. There are two ways of running a python script; either runs a script from the python interactive shell or from the IDLE. Let us say you want to run a file called **names.py** in IDLE. You will begin by opening IDLE in the main interactive mode by pressing **Ctrl + O** or select and open the file menu. Once the opening window shows up, scroll to the location of the file you want to run (in this case, the **names.py** is stored in the directory **/home/pi/py3prog.** Select the script by tapping on it and then choose open to have access to the file.

When you tap the open tab, you will be greeted with another IDLE window, which will display the python script with its location and name as **names.py-/home/pi/py3prog.** You will now be able to run the python script right from its window by pressing on the F5 key or select **run** from the menu bar and then run the program.

Creating a python script in IDLE

To create your python script in IDLE, start by opening the text editor from the IDLE interactive mode window by tapping **Ctrl + N.** You can alternatively select the file menu and choose a new window. A new window will be prompted with the inscription "untitled" on the title bar space. This brings you to the basic IDLE text editor. In the basic IDLE text editor, type in the Python statements to create your script. When you are all done, you save the statements to a file. If you want to save the script to a file, simply press **Ctrl+ S** or tap on the file menu and choose save. You will be prompted with a "save as" window where you will be able to select the directory where you need the file to be saved. Enter the name of the file and hit the save button. All python programs are named as **filename.py.**

One popular text editor, which is usually available by default in the Raspbian OS, is the Nano text editor. The nano text editor is ideal for the Raspberry pi because of its lightweight feature, and it is also easy to use when compared with the other text editors. Nano's biggest strength over most other text editors is that it can be deployed in the graphic user interface and the command line. Its biggest advantage over the text editor in IDLE is that nano can be used in both the GUI and at the command line! To use the nano text editor at the command line, you just type the command nano and press Enter. To start the nano text editor in the GUI, simply tap on the LXDE Programs Menu icon located at the far left of the LXPanel. Place the cursor over the other menu so that you can access its submenu, and then select nano menu. Note that the nano text editor is not able to carry out any syntax checking while you write in Python statements. It also does not do any color-coding while you write statements. And it does not perform any auto-indentation. nano doesn't give you any handholding when you're creating and editing Python scripts. The Nano text editor will be used within the graphic user interface through the LXTerminal. The title bar of the nano editor program window is the line where the left side starts with "GNU nano" and the nano editor version number. In the middle of the title bar are either the words "New Buffer" if you are creating a new file or the name of the file you are editing.

Understanding the Python Basics

Producing python script output

As a beginner, you need to know how to produce output from a python script. The **print function** is used for this purpose. When you put a group of python statements together to perform a defined task, they are called **functions**. The **print** became a function with the coming of the python v3. The function of the **print** is to produce items as output. These items are called **arguments**. Therefore, the syntax of the **print function** is *print (argument)*. The *argument* of the **print function** can be character (such as letters or numbers) and can also be a variable.

To use character as print function argument: characters are shown using a set of single quotes or double quotes. For single and double quote, the result will not contain the quote as depicted below;

>>> **print ('The prince received a gold medal')**
The prince received a gold medal.
>>>

>>> **print ("The prince received a gold medal")**
The prince received a gold medal.
>>>

If you decide to deploy single quotation marks to enclose characters in a print function argument, then use the single quote all through without mixing it with double quotes. Although python does not care whether you mix them or not, however, it is poor formatting if you deploy a single quote in one print argument and then use a double quote in the other ones.

Note that when you have a statement that already has a quote inside (showing possession), endeavor to use a double quote in the print argument. For instance;

>>> **print ("The king's daughter is married")**

The king's daughter is married.

>>>

Output control with Escape sequences

The escape sequence contains a series of characters that enable a Python statement to "escape" from its ordinary behavior. The new behavior might be attributed to the inclusion of some special formatting or the protection of some special characters or functions that you are using in syntax. Escape sequences all start with a backslash (\) character. A typical instance of deploying an escape sequence to include special formatting is the \n escape sequence. The \n escape sequence takes any characters you placed after it onto the next line of the output. This is referred to as a **newline,** and it inserts a line feed formatting character. See below a typical instance of using the \n sequence to include a linefield;

>>> **print ("physics is a branch of science.\nbiology is harder than chemistry.\nAnd farming is good.")**
   Physics is a branch of science.
   Biology is harder than chemistry.
   And farming is good.
>>>

With the Unicode escape sequence, you will be able to use all types of characters with your output.
You will be able to show Unicode characters by deploying the \u escape sequence. Each Unicode character is marked by a hexadecimal number. You can navigate the Unicode website at www.unicode.org/charts to see more of the hexadecimal numbers. There are many Unicode characters which are anchored by the

hexadecimal numbers. The pi (Π) symbol has a hexadecimal number of 03c0. You can show the pi (Π) symbol using the Unicode escape sequence by starting the number with a \u in the print function argument. See an example below;

>>> **print ("I know a lot about the Raspberry \u03c0!")**
    I know a lot about Raspberry π!
>>>

Formatting scripts for readability

Sometimes, you may need to show a longer output line by using the print function, perhaps some line of instructions that you need to send to your script user. The longer output lines are not advisable as they render the code harder to understand with poor readability. You need to break down these long lines of outputs. You can do this in a couple of ways;

The first approach that you can use to truncate a long line of characters is to actually deploy **string concatenation.** The string concatenation will bring two or more units of text and combine them together so that they will turn to one text string. The "combine" here is a "+" that will be used. To make this to work perfectly, you need to escape out of the ordinary print function by using the escape sequence. You can do this by deploying the backslash (\) symbol. This will escape you out of the usual print function of inserting a line feed at the end of a string of characters. Therefore, the two things you will need to truncate the longer output line are "+" and a backslash (\). See example below;

```
>>> print ("There were times when the economy was good" +\
... "that we used to buy goods at cheaper prices!")
There were times when the economy was good that we used to
buy goods at cheaper prices!
>>>
```

It is obvious in the instruction above that the two strings have been concatenated as a single string in the result. There is another method of doing this, which is even simpler and straightforward. In this method, you will not include the "+" and the backlash symbol (\). You will only need to maintain each character string in its set of quotation marks. The character will be concatenated by the string function automatically. See the example below;

```
>>> print ("There were times when the economy was good" ...
"that we used to buy goods at cheaper prices!")
There were times when the economy was good that we used to
buy goods at cheaper prices!
>>>
```

Notice the removal of the +/ sign in the above example.

Writing comments in scripts

The author of a python script can embed a note in the script. Notes like that are referred to as comments. The author's intention, most of the time, for putting comments in scripts is to give further clarification concerning the syntax and logic of the script. The python language doesn't, particularly, make use of these comments for anything and it does not affect the script whatsoever. These comments are only useful to people who are making use of the script to either edit or debug the script. If you want to add comments to your script, you start the comment with a pound or hash (#) sign. The python language will not reckon with the texts that come after the hash sign. As the author of a script, it is in your best interest to add your name, the reason why you wrote the script and the time and date the script was written. Although, these types of information (You Name, the reason why you wrote the script and the time and date the script was written) can be, sometimes, placed at the top of your scripts or at the bottom as the case may be. Nevertheless, inserting your name and identity as a comment is a good way of getting credit for your work when people share them across platforms.

Understanding the python variable

A variable can be likened to a name that secures a value for future use. While naming python variables, you need to be aware that variable names are case sensitive. For instance, variable names CarbonPrint and Carbonprint are not the same. Even though they have the same spelling and look alike, they are different when it comes to python because the **print** in the first one started with a capital letter while a small letter was used in the second one. The following rules are especially useful when creating variable for python language;

- **Do not use a python keyword as variable name:** Python updates its list of keywords very often. So, it is expedient that you update yourself with a list of current python's keywords. You don't need to commit these to memory as you can always take a look at the update anytime you need them. To check the keywords, you will need to deploy a function that is part of the standard library. However, this function is not necessarily a built –in function like the print function. This function is already available on your Raspbian OS. But you need to call the function into python if you want to use it. The name of the function is **keyword**. Check below to see how to import into python and know the keyword.

>>> **import keyword**
>>> **print (keyword.kwlist)**
['False', 'None', 'True', 'and', 'as', 'assert', 'break', 'class', 'continue', 'def', 'del', 'elif', 'else', 'except', 'finally', 'for', 'from', 'global', 'if', 'import', 'in', 'is', 'lambda', 'nonlocal', 'not', 'or', 'pass', 'raise', 'return', 'try', 'while', 'with', 'yield']
>>>

The **"import keyword"** brings the keyword functions for use in the python interpreter. Then the prompt **print (keyword.kwlist)** deploys the keyword and print function to show you the list of current python's keywords. You should not use these keywords as python variable.

- **Do not use a number as the first character of your variable name**: the first character in the variable name can be any letter from "a" to "z", any letter from "A" to "Z" and an underscore (_). After the first character, you can deploy any number from 0 through 9. Since you cannot use space between a variable name, you can actually use an underscore in between the name for improved readability.

Assigning value to a variable

It is very easy to assign value to your variable in python. What you do is write the variable name first, and then put an "equal to =" in front of the name, after which you end it with the value of the variable. The syntax is **variable = value.**

Formatting variable and string output

When you use variables, you automatically add extra formatting issues to the prompt. For instance, the print function will add a space anytime it sees a comma in a sentence or statement. This explains why you should not add any extra space at the end of the text. Sometimes, you might want to use another character aside from space to space out a string of character from a variable in the result. In a case like this, you can use a separator in the statement. The example below uses a *Sep* to put an asterisk (*) in the statement rather than a space;

```
>>> tea_cup = 'tea'
>>> print ("I love my", tea_cup, "!", sep='*')
    I love my*tea*!
>>>
```

Assigning long string values to variables

If you want to assign a long string value to your variables, you can divide the string into many lines by using many methods. One of the methods involves deploying string concatenation, a + symbol, to combine the strings together and the escape character (\) to prevent a line feed from being added. See the example below where two different lines of texts were concatenated in the assignment of the variable crude_oil;

>>> **crude_oil = "America has a large reserve of crude that can last eternity" +**
... **"and provides basic energy requirement for the whole country!"**
>>> **print (crude_oil)**
    America has a large reserve of crude that can last eternity and provides basic energy requirement for the whole country!
>>>

Another way is to utilize parentheses to enclose the variable's value. The example below has totally eliminates the +\ and deploy parentheses on both side of the whole long string so as to make it becomes a single long character string;

>>> **crude_oil = ("America has a large reserve of crude that can last eternity"**
... **"and provides basic energy requirement for the whole country!")**
>>> **print (crude_oil)**
    America has a large reserve of crude that can last eternity and provides basic energy requirement for the whole country!
>>>

Assigning numeric values to variables

It is not only a string of characters that you can assign to your variables; you can as well assign numbers to your variables. For instance, if we assign variable cups_consumed to the number of cups of tea consumed, we can have strings like the example below;

>>> **tea_cup = 'tea'**
>>> **cups_consumed = 5**
>>> **print ("I had", cups_consumed, "cups of",**
... **coffee_cup, "today!")**
    I had 5 cups of coffee today!
>>>

You can, at the same time, assign the output of an expression to a variable. The equation 5+1 is completed in the example below, and then the value 6 is assigned to the variable cups_consumed.

>>> **coffee_cup = 'coffee'**
>>> **cups_consumed = 5 + 1**
>>> **print ("I had", cups_consumed, "cups of",**
... **coffee_cup, "today!")**
    I had 6 cups of coffee today!
>>>

Reassigning a variable

The value you assigned to a variable is not its birth right, as the value can be switched to be assigned to another variable. That is why it is called variable in the first place. The variable coffee_cup can have its value switched from coffee to butter. You can reassign a value by entering the assignment syntax with an entirely new value at the end.

```
>>> coffee_cup = 'coffee'
>>> print ("My cup is full of", coffee_cup)
    My cup is full of coffee
>>> coffee_cup = 'tea'
>>> print ("My cup is full of", coffee_cup)
    My cup is full of tea
>>>
```

Python data type

Whenever you create a variable by an assignment, like **variable = value,** the python language will help you to determine the data type, and then assign the data type to the variable. When a variable is created by an assignment such as *variable = value*, Python determines and assigns a data type to the variable. A *data type* is the one that determines how you are storing your variables and also provides rules that guide how you can manipulate the data. Python utilizes the value assigned to a variable to know the data type. When you entered the Python statement coffee_cup = 'tea', Python observed the characters in quotation marks and determined the variable coffee_cup to be a ***string literal*** data type, or *str*. The **type** function can be used to know the data type that python has assigned to a variable. For instance, in the example below the variable **coffee_cup** has been assigned the data type **str** because python saw a string of characters between the quotation marks. While the

variable **cups_consumed** has been assigned **integer int** because python saw a whole number;
```
>>> coffee_cup = 'coffee'
>>> type (coffee_cup)
    <class 'str'>
>>> cups_consumed = 3
>>> type (cups_consumed)
    <class 'int'>
>>>
```

If you make a small modification in the **cups_consumed** assignment statement, python will automatically change the data type. For instance, if you reassigned the number assigned to cups_consumed, let us say from 3 to 3.5, python will change the data type assigned to cups_consumed from integer to float data type.
```
>>> cups_consumed = 3
>>> type (cups_consumed)
    <class 'int'>
>>> cups_consumed = 3.5
>>> type (cups_consumed)
    <class 'float'>
>>>
```

Allowing python script input

Sometimes, you may require your script user to give answers to some questions or give you some data inside your script. Python leverages the **input function** to get this done. The input function is essentially a built-in function with the syntax; **variable = input**

(user prompt). In the example below, the input function returned the value assigned to the variable cups_consumed. The user of the script was asked to give this information. The prompt that the script developer provided to the script user is indicated inside the input function as an argument. The script user just needs to give the answer and hit the Enter button. This automatically enabled the input function to assign the value 3 to the cups-consumed variable.

```
>>> cups_consumed = input("How many cups did you drink? ")
    How many cups did you drink? 3
>>> print ("You drank", cups_consumed, "cups!")
    You drank 3 cups!
>>>
```

The input function sees all input as strings. This is totally different from the way python treats other variable assignments. Note that if cups-consumed = 3 was inside your script, python would assign it the integer (int) data type. The input function in the example below is set to string (str);

```
>>> cups_consumed = 3
>>> type (cups_consumed)
    <class 'int'>
>>> cups_consumed = input("How many cups did you drink? ")
How many cups did you drink? 3
>>> type (cups_consumed)
    <class 'str'>
>>>
```

You can change variables (input from the keyboard) from strings by using the integer function. The integer function will switch a number or value from a string data type into an integer data type. You can as well use the float function to switch functions from a string data type into the float data type. The example below switch variable cups-consumed from string to an integer data format;

\>>> **cups_consumed = input ("How many cups did you drink? ")**
How many cups did you drink? **3**
\>>> **type (cups_consumed)**
    <class 'str'>
\>>> **cups_consumed = int(cups_consumed)**
\>>> **type (cups_consumed)**
    <class 'int'>
\>>>

When you use a **nested function** with the input, it makes your script more concise and straight to the point. The only disadvantage is that your script might be a bit uneasy to read.

\>>> **cups_consumed = int(input("How many cups did you drink? "))**
How many cups did you drink? **3**
\>>> **type (cups_consumed)**
    <class 'int'>
\>>>

Working with numbers and performing calculations in the Raspberry python script

Python Math operators

You can confirm if python supports all the basic Math operators by opening the IDLE window and performing some random calculations. Using the IDLE command prompt, you can input any type of mathematical calculation or equation and the window will help you to evaluate it. For instance, the following basic calculation can be entered into the IDLE window for experimentation;

```
>>> 3 + 1
4
>>> 5 - 4
1
>>> 2*8
16
>>> 18 / 9
2.0
>>>
```

For the division task, python converts the result automatically to float data type even when the mathematical inputs are both integer. This is a new thing in the python v3.2

The following mathematical operators are also supported by python in your script;

// (floor division), ** (exponentiation), & (binary AND), | (binary OR), << (binary shift left), >> (binary shift right), and (logical AND), or (logical OR), not (logical NOT) and modulus.

Order of operation

Just like in every mathematical calculation that involves BODMAS which provides a template for simplifying addition, subtraction, multiplication and all other functions. The python programming language also follows rules while solving mathematical problems involving addition, multipli-cation and subtraction. In the example below, python carry out the multiplication operation first before the addition operation;

>>> 7 + 3 * 2

13

>>>

But the order of operation above can be changed with the help of parentheses;

>>> (7 + 3) * 2

20

>>>

Using variables in calculation

One of the most fascinating features of using math in python is that it enables you to place variables inside your equation. If you mix data type in your equation, python will display the result in form of floating-point data type;

>>> test1 = 7

```
>>> test1 * 3.0
21.0
>>>
```

You need to assign a value to a variable before you deploy it in your calculation, otherwise python won't agree. See the example below;

```
>>> test1 * 5
Traceback (most recent call last):
File "<pyshell#0>", line 1, in <module>
test1 * 5
NameError: name 'test1' is not defined
>>>
```

While working with python calculation, you must have observed some strangeness in the floating point outputs. See below;

```
>>> 5.2 * 9
    46.800000000000004
>>>
```

When you, ordinarily, multiplied 5.2 by 9, the result should be 46.8. But the IDLE displays the output with some stray numbers added to it. This is due to the way the CPU computes floating-point calculation. Because the floating point data type changes the numbers into a special format, the calculations are somewhat inaccurate. This problem is difficult to by-pass in your calculation but you can deploy some tricks to reduce the extent of the stray values and make your calculation presentable. You can deploy the format function inside the result to modify how the result is being displayed. The format () function enables you to separate the variable from the output text in the string object just by using the {} placeholder symbol;

>>> print ("The result is {}".format (result))
    The result is 46.800000000000004
>>>

You can then use the placeholder symbol {} to further format your result. All that you need to do is to specify an output template inside the placeholder {} and python will deploy it to format the number output. For instance, you can restrict the output of the calculation above to two decimal places by using the template {0:.2f} to produce the output:

>>> print ("The result is {0:.2f}".format (result))
    The result is 46.80
>>>

The first number in the above template tells python what position in the number it should start to display from. The second number (the .2) tells python the number of decimal places it should include in the output. The f in the template informs Python that the number is actually a floating-point format.

Creating complex numbers in python

In python, you can specify complex numbers by deploying the complex function (). To get started, indicate the real component digit as the first parameter and then put the imaginary component as

the second parameter. For instance, if you have complex function 2 + 5j, the real part is 2 while the imaginary part is 5. See example below;
```
>>> test = complex(2, 5)
>>> print (test)
    (2+5j)
>>>
```
Python will usually display the complex number value using the j format which makes it look presentable and easy to view.

It is possible to carry out complex number calculation within another complex number. See below;
```
>>> test1 = complex (1, 2)
>>> test2 = complex (2, 3)
    >>> result = test1 + test2
>>> print (result)
    (3+5j)
>>> result2 = test1 * test2
    >>> print (result2)
(-4+7j)
>>>
```

Using the NumPy Math Libraries

Although, there are many mathematical functions which are of diverse importance in the standard math module - most scientists, researchers, engineers and statisticians love to create and deploy their own advanced extended math modules. A very powerful python library for advanced mathematical computation is NumPy. The NumPy module gives templates for multidimensional array manipulations which are of importance in many advanced scientific and engineering calculations. The NumPy is made up of the following;

A multidimensional array object class

Methods for array manipulation

The NumPy multidimensional array objects are, to some degree, different from the standard Python lists or tuples because you can easily deploy them in many mathematical calculations that need arrays. The way python takes care of the array object differs essentially from the way it does take care of lists and tuples.

NumPy data type: The NumPy module is made up of five main data types that you can find applicable to store your data in arrays. They include; **bool**—Booleans, **int**—Integers, **uint**—Unsigned integers, **float**—Floating-point numbers and **complex**—Complex numbers.

It is easy to include a bit size at the end of the data type name. Bit size such as float64, int8, or complex128 can be specified. If you fail to specify the bit size, python will automatically assume your bit type based on the CPU platform.

To start using the NumPy functions in your work, start by importing the NumPy module. The NumPy module might take a little time to load all of the libraries so you need to exercise some patience.

Creating NumPy Arrays

When you install the Raspberry pi python version 3, you automatically get the NumPy modules installed with it. You can begin to write your advanced array manipulations without a hitch. There are many ways you can create arrays in NumPy. One method is to simply create the array from the existing python list or tuple. See the example below which creates a 3 by 3 array by deploying three python list;

```
>>> import numpy
>>> a = numpy.array(([1, 2, 3], [0, 2, 4], [3, 2, 1]))
>>> print(a)
    [[1 2 3]
     [0 2 4]
     [3 2 1]]
>>>
```

You need to specify the data type; otherwise python will assume a data type for you. Whenever you need to change the data type of your array value, you should specify it as a second parameter to the array () function. The example below makes the values to be stored in the floating data type;

```
>>> a = numpy.array(([1,2,3], [4,5,6]), dtype="float")
>>> print(a)
[[ 1. 2. 3.]
 [ 4. 5. 6.]]
>>>
```

You can as well generate an all zero or all one default array like the one below;

```
>>> x = numpy.zeros((3,5))
>>> print(x)
[[ 0. 0. 0. 0. 0.]
 [ 0. 0. 0. 0. 0.]
 [ 0. 0. 0. 0. 0.]]

>>> y = numpy.ones((5,2))
>>> print(y)
[[ 1. 1.]
 [ 1. 1.]
 [ 1. 1.]
 [ 1. 1.]
 [ 1. 1.]]
>>>
```

Using the "if" statement

The **if** statement is the most basic form of structured command. In python, the **if** statement has the formatThe most basic type of structured command is the if statement. The if statement in Python has the following basic format:

if (*condition*): *statement*

Other programming languages use the "then" keyword in their "**if**" statement, but the python programming language has replaced the "then" keyword with a semicolon. Python examines and then solves the condition inside the parentheses and can either execute the statement immediately after the semicolon (if the condition returns a True logic value) or skips the statement after the semicolon (if the condition returns a False logic value).

Let us walk through the "do it yourself" example below to familiarize you with the "**if**" statement.

- **Open your Python 3 IDLE interface located right on your graphical desktop.**
- **Set a value for a variable**:

>>> a = 40

- **Test the value of the variable by using an if statement:**

>>> if (a == 40): print("The value is 40")
The value is 40.

- **Try another test using another condition:**

>>> if (a < 100): print("The value is less than 100")
The value is less than 100

- **Try another test condition that will fail**:

```
>>> if (a > 100): print("The value is more than 100")
>>>
```

When you are using the "**if**" statement, any time you input the statement and then hit the Enter key, the IDLE platform will pause on the following line to confirm if you are still going to input any more statements. You can just hit the Enter key once again to close out the statement.

In the first script, the condition checks to confirm if the variable "a" is equal to 40. Since it is equal to 40, Python prompts the print () statement on the line and prints the string. In the same vein, the second script confirms whether the value stored in the "a" variable is less than 100.

Since it is actually less than 100, Python will execute the print () statement again to display the string. But in the third case, the value stored in the "a" variable is not greater than 100, so the

Condition sends back a False logic value, making Python to skip the print () statement after the semicolon.

Using the "else" statement

The "if" statement will only run your statement if the condition is true. But if the condition sends back a false logic value, the python programming language will take a swift swerve and move to the next statement. In order to run the statement even when the condition returns a false logic value, you need another tool. This is where the "else" statement comes into play. The "else" statement will initiate another group of commands in the statement;

```
>>> x = 25
>>> if (x == 50):
print("The value is 50")
else:
    print("The value is not 50")
The value is not 50
>>>
```

When you combine the else statement together with the "if" statement, you need to be careful how you place the else statement. If you try to keep it indented, you will receive an error message from Python:

```
>>> if (x == 50):
    print("The value is 50")
    else:
SyntaxError: invalid syntax
>>>
```

The same case is applicable when you are using the "if" and else statements in Python scripts. When you are initiating your script code file, ensure that you line up the else statement properly in the text.

Learning about Loop in python

Using loop to perform repetitive tasks

Your computer is able to perform one task over and over again and doesn't get bored like a human being. This process of carrying out a task over and over again is called **repetition.** Another word for **repetition** is **iteration.** When it comes to programming language, **iteration** is the act of carrying out a set of tasks repeatedly until a desired result is obtained or the task has been done for a desired length of period. **Iteration** in python, essentially, refers to loop. One time through a loop is referred to as **one iteration.** When you go through a loop multiple times, you refer to it as **"iterating through the loop".**

Using the "for loop" for iteration

In Python programming language, the "for loop" construct is known as a "count-controlled" loop. This is because the loop's set of activities will be carried out a set number of times. Let us say you want a set of actions to be done six times, you can deploy a "for loop" in python to help you carry out this task. The syntax structure of the "for loop" in Python is as follows:

    for *variable* in *data_list*:
        set_of_Python_statements

You should observe that there is no ending statement in the "for loop." You will see a "done" or "end" type of statement. In a "for loop," the python statements which are to be included are usually

indented under the "for" construct. The operation of a "for loop" is as follows:
- The *variable* in the "for" construct is given the first value in the data list.
- The Python statement(s) in the loop will then be executed and has the liberty of using the variable value that was assigned to it.
- As soon as the loop's python statement is completed, the variable will be reassigned the next value in the data list.
- The Python statement(s) in the loop will then be executed and has the liberty of using the variable value that was reassigned to it.
- The "for loop" will continue until you have assigned all the values to the variable and the python statement(s) executed during each assignment of the variable.

Validating a user input with a "for loop"

You can obtain input from people using your script (which has already been explained in the previous concepts) and then validating the input. This is referred to as **input verification.** You can write a script where you can allow the script users three attempts to enter the correct input. In the same vein, you can try something new by inserting a break. Kindly follow these steps;
- Boot on your Raspberry pi by connecting it to the power source and log into the system.

- Start the LXDE GUI by prompting the command **startx** and then press Enter. Open the LXTerminal by double-clicking the LXTerminal icon.
- The command line prompt will be displayed where you will need to type in **nano py3prog/script0701.py** and then hit the Enter button. The command puts you into the nano text editor where you will be able to create the file **py3prog/script0701.py.**
- Enter the following codes into the nano editor window and press the Enter key at the end of each line:

```
# script0701.py - The Secret Word Validation.
# Written by <your name here>
# Date: <today's date>
#
############## Define Variables ##########
#
# max_attempts =3              #Number of allowed input attempts.

the_word = 'secret'            #The secret word.

#
############### Get Secret Word ################
# for attempt_number in range (1, max_attempts +1):
secret_word = input("What is the secret word?")
```

```
if secret_word == the_word:
    print ()
    print ("Congratulations! You know the secret word!")
    print ()
    break             # Stops the scripts execution.
else:
    print ()
    print ("That is not the secret word.")
    print ("You have", max_attempts - attempt_number, "attempts left.")
    print ()
```

- Write out the modified script you just typed above inside the text editor by hitting Ctrl+O. Click Enter to write out the contents to the **script0701.py script**.
- Exit the nano text editor by hitting Ctrl+X.
- Write **python3 py3prog/script0701.py** and hit Enter to run the script. Answer the question correctly the first time you run the script by entering a secret.
- The script stops once you input the correct answer due to the break statement. The break statement prompts the loop to terminate. In other words, the break statement lets you "breakout" of the loop.

- Type **python3 py3prog/script0701.py** once again and hit Enter to run the script. This time, answer the question incorrectly, answering anything except secret.

Using the while loop for iteration

The while loop construct, in python, is referred to as "condition-control" loop. This is because a desired condition must be established before the loop's set of tasks are carried out. The iteration will stop immediately when the condition is met. For instance, you might need a set of loop's tasks to be carried out until a specific condition is no longer true. In such instances, you need a while loop. The syntax structure of the while loop in Python is as follows:

while *condition_test_statement*:
    set_of_Python_statements

Just like the case in the "for loop," the while loop deploys indentation to denote the specific python statement that is linked with it (code block).Just like the "for loop," the while loop uses indentation to denote the Python statements associated with it (code block). *condition_test_statement* examines a condition, and if it finds the condition to be true, it will execute the python statement within the loop's code block. The condition for each iteration is checked, and the iterations stop once the condition is examined to be false.

A loop will come in handy when you are thinking of entering data. You can create a script to enter a club's member list with a while loop. The script will ask for the number of the member names you want to enter. The while loop will request for the member's identity, such as their surname, first name and middle name. Follow the steps below to get started.

- Boot on your Raspberry pi by connecting it to the power source and log into the system.
- Start the LXDE GUI by prompting the command **startx** and then press Enter.
- Double click the IDLE 3 icon to open the IDLE 3 window.
- Tap ctrl + N to open the text editor window.
- Enter the code shown in the image below inside the IDLE text editor window. Tap Enter when you need to get to the next line.

```
# script 0702.py - Enter python Club Numbers using While loop
# written by <your name here>
# Date: <Today's date>
#
################## Define variables ################
Names_to_enter = int (input ("how many python club member names to enter?"))
Name_entered = 0
#
While   names_to_enter > names_entered:
        # iterate till all names entered
                Member_Number = names_entered + 1
                Print ()
                Print ("number + 1" + str (member_number))
                First_name = input ("first name: ")
                Middle_name = input ("Midle name: ")
                Last_name = input ("Last name: ")
                Names_entered += 1
                Print ()
                Print ("member #" member_number "is," first_name, middle_name, last_name)
```

- Test your new scripts in IDLE by tapping the F5 key and then input the answers to the questions.
- You can save the script by tapping ctrl + S to have access to the "save as" window. Click the py3prog folder icon twice and then enter script0702.py inside the file name bar. Tap the save button to save.
- Tap on the Ctrl + Q to leave the IDLE environment.

Creating and using Tuple in programming

The tuple is a category of data type in Python which enables you to store multiple data values that are constant (doesn't change). In the programming community, these data values are considered to be *immutable*. Once you create a tuple, you can either work with the tuple as a single object or you can reference the individual data value inside the tuple in your Python script code. In python, you can create tuple using four methods and these include;

- Use parentheses to input an empty tuple value. See below

```
>>> tuple1 = ()
>>> print(tuple1)
()
>>>
```

- Insert a comma immediately after a value in an assignment. See below:

```
>>> tuple2 = 1,
>>> print(tuple2)
(1,)
>>>
```

- Use commas to separate multiple data in an assignment. See below;

```
>>> tuple3 = 1, 2, 3, 4
>>> print(tuple3)
(1, 2, 3, 4)
>>>
```

- Specify an iterable value by using the tuple () built-in function in python. See below;

```
>>> list1 = [1, 2, 3, 4]
>>> print(list1)
[1, 2, 3, 4]
>>> tuple4 = tuple(list1)
>>> print(tuple4)
(1, 2, 3, 4)
>>>
```

Creating and using lists in python

Just like tuples, lists store multiple data values which are referenced by a single list variable. One difference between tuples and lists is that: tuples are immutable while lists are mutable (you can change or delete any data variable you stored in lists).

Just like it is with tuples, there are four ways of creating a list;
- Use an empty pair of square brackets to create an empty list. See below;
```
>>> list1 = []
>>> print(list1)
```

[]
>>>

- Insert square brackets around a list of values that are separated by commas. See below; >>> list2 = [1, 2, 3, 4]

>>> print(list2)
[1, 2, 3, 4]
>>>

- Create a list from any other iterable object by using the list function (). See below;
>>> tuple11 = 1, 2, 3, 4
>>> list3 = list(tuple11)
>>> print(list3)
[1, 2, 3, 4]
>>>

- Use a list comprehension.

The list comprehension method gives you a genuine way of creating a list from other data. You can observe that python uses square brackets in the data value which are different from the parentheses used by tuples. Like it is in tuples, lists can have any type of data, including numbers and any other data type. See below;
>>> list4 = ['John', 'Mercy', 'Tim James', 'Torres']
>>> print(list4)
['John', 'Mercy', 'Tim James', 'Torres']
>>>

CHAPTER FIVE

USING PROGRAMMING IN THE RASPBERRY PI

The Graphical User Interface (GUI) Programming

The graphical user interface (GUI) on all operating systems enables you to enter data and see the processed result in a jiffy. Like most other operating systems, the Linux OS – which is the building block of your Raspberry pi – also has its own graphical user interface. There are many graphical desktop environments in the Linux OS, but the Raspbian OS which is used by the Raspberry pi deploys the LXDE desktop environment to give you an appealing graphical desktop interface. You can use the LXDE desktop environment together with python script to invent a more appealing window-oriented interface that can give your programming work a very good look. You can go through some terms used in the graphical user interface programming first;

- **Frame**: this is the main section in a window. It is usually made up of objects called **widgets.**

- **Widgets:** helps display and retrieve information. The following are the types of widgets you might come across with their functions;
    - **Frame:** provide a window section that you can utilize to put other widgets inside the window.
    - **Label:** allows you to write texts in the window section.
    - **Button:** click a button to prompt an event in the window interface.
    - **Checkbutton:** allows you to select or unselect an item.
    - **Entry:** provide a section to write or show a single line of texts.
    - **Listbox:** gives you an array of values to choose from
    - **Menu:** gives the menu toolbar at the window top section
    - **Progressbar:** let you know when an action is taking place in the background. When you are downloading something or copying a file, the progress button will usually show to tell you the extent of the process.
    - **Radiobutton:** enables you to choose on item from a list of items
    - **Scrollbar:** manages the view in a frame
    - **Separator:** put a horizontal or a vertical bar inside the window.
    - **Spinbox:** enables you to choose a value from an array of numbers.
    - **Texts:** an area to write multiple lines of texts.

A GUI program shows all the widgets at once inside the window. It is the user that will get to determine which widget he wants to activate next since the code cannot determine which widget the user is planning to use at any point in time. The GUI program deploys a technique called *event-driven programming* to process codes. In event-driven programming, Python initiates different methods inside the program, using the series of actions that has happened in the GUI window. The program flow doesn't exactly have a set flow; it only runs in response to a triggered event. For instance, your user can input their data or information into the text widget, and nothing will happen unless the user taps a button in the program window to submit the text. The tapped button initiates an event, and your program code needs to detect that event and then prompt the code method to read the text in the text field and then start processing it.

The basis of the event-driven programming is linking widgets inside the window to some specific events and then linking these events to the code modules inside the program. The *event handler* is the one carrying out this process.

Python GUI packages

There are many excellent GUI packages that will enable you to create GUI widgets using your python scripts. These widgets will allow you to create your own graphical program. The most common GUI packages include; **tkinter, PyGTK, PyQT and wxPython**.

The python programming language features the **tkinter** package by default, hence it is mostly used to create python's graphical interface on your Raspberry pi. The **tkinter** is also among the oldest graphical packages in python programming.

The tkinter package

The library of the python Raspberry pi features the **tkinter** package by default, so it is mostly used to show how to create graphical user interface programs in the python script. There are three (3) essential steps you need to be familiar with in order to create a graphical user interface application using the **tkinter** package. These steps include;

1. Creating a window.
2. Adding widgets to the window.
3. Defining the event handlers for the created widgets.

- **Creating a window:** The first thing you need to do before you create a graphical user interface program is to set up the main window for your application. This main window will be called the **root window** and it can be achieved by creating a Tk object. If you want to create a Tk object, import the tkinter library first and then place a Tk object just like it is done below;

*from tkinter import ***
root = Tk()

The entry above will create a main window object and assigns the main window object to the variable named root. However, the default window you set up above will lack most of the window features such as title, size or any other essential features. You need to set up some of these window features by running a few defaults Tk objects. You can run the **title method ()** which will assign the title for the window. This title will be displayed at the top part of the window. You as well set up the **geometry method ()** that will assign size to the window. See below;

root.title('This is a test window')

root.geometry('300x100')

As soon as you are done setting the title and the size for the window, you will need to place the window into a loop by deploying the **mainloop ()** method. This will make the window gadget to trigger an event. Python will be able to intercept any event that happens in the window and then passes these events into the program code. For instance, if you tap the X menu at the top right section of the window, python will record this action and will understand that you intend to close the window. Check the tkinter window code below for **script1801.py** to create a simple window.

#!/usr/bin/python3

from tkinter import *

```
root= Tk()
root.title('This is a test window')
root.geometry('300x100')
root.mainloop()
```

You need to enter into the LXDE graphical desktop on your Raspberry pi in order to run and execute the **script1801.py** code. When you enter the LXDE graphical desktop on the Raspberry pi, simply initiate the script right from the command line by clicking to open the LXTerminal menu. Run the code from the command prompt of the LXTerminal menu. See the code below;

```
pi@raspberrypi ~$ python3 script1801.py
```

- **Adding Widgets to the window:** Once you have successfully created the root window, you can now start adding widgets to the window interface. The following three (3) steps are involved when you want to add widgets to your window;

    1. Creating a frame template in the root window.
    2. Defining a positioning method you want to use for placing your widgets in the frame.
    3. Placing the widgets in the frame by deploying the positioning method you have defined.

- **Creating a frame template in the root window:** This is mainly the first step involved in adding widgets to your window. This step involved creating a template for the window widget layout. The Frame object which has been customized with the **tkinter** package will be deployed to create a frame area for you to put your widgets. However, the frame area cannot be used right in your window code. You need to create a child class that will define all the window methods and attributes using the frame class. You can use any name of your choice for the frame object child class. However, the most commonly used name for the frame object child class is **Application.** See below;

class Application(Frame):

Creating the child class is not enough as you need to create a constructor with the __init__() method. The __init__() method deploys the keyword self as its first parameter, and it uses the root Tk window object which you have already created as its second parameter. The first and the second parameter is the one that associates the Frame object to the window. With this, you now have a readymade template that you can use to create your window Frame class. See below;

class Application(Frame):

"""My window application"""

def __init__(self, master):

```
super(Application, self).__init__(master)
self.grid()
```

The method highlighted above will create the window using the bare Tk object. If you now want to add a frame so that you can start arranging your widgets, see the below **script1802.py;**

```
#!/usr/bin/python3
from tkinter import *
class Application(Frame):
    """Build the basic window frame template"""
    def __init__(self, master):
        super(Application, self).__init__(master)
        self.grid()
root = Tk()
root.title('Test Application window')
root.geometry('300x100')
app = Application(root)
app.mainloop()
```

- **Positioning your widgets:** In order to have a user-intuitive Graphical User Interface application, you need to understand

and get familiar with the right way of adding widgets in the window area. You don't need to cluster too many widgets together in the window area as it makes the whole interface messy.

From the previous section, you were able to use the grid method to arrange your widgets in the frame. Now, with the **tkinter** package, you can use the three methods below to arrange your widgets inside the window;

    Using a grid system

    Arranging widgets into available places

    Using positional values

The third method, which deploys the positional values of arranging widgets in the window, will need you to specify the exact location of each widget you are adding by using the X and Y axis within the window. This method might be difficult for you as a beginner, though it gives a perfect way to control where you are placing your widgets. The packing method tries its best and it is used to pack widgets, as much as it can, into the window inside the space available. The packing method will allow window to put your widget in the window beginning from the top left and then move along to the next available space, either to the right of the previous widget or below the previous widget. The packing method works best if you have a small window with not so many widgets to arrange. The packing method gets messy if you are working with a larger window and make your widget get cluttered and in bad view. The judge between the packing method and the positional method is the grid method. The grid method affords rows and columns to create a grid in the window. Widgets are placed across rows and columns. You can define a widget with the grid method using the three parameters below;

object.grid(row = x, column = y, sticky = n)

The third parameter, which is the sticky parameter, tells the program to align the widget inside the row and column. The sticky parameter has nine sticky values which define the orientation and alignment of each widget using the four cardinal points (**N, S, W, and E.**), two combinations of the cardinal points (**NW, NE, SW, and SE.**) and the center alignment. Each of these sticky values places the widget at the different location and direction in the window. For instance, specifying the **N** will place the widget at the top of the cell and vice-versa.

Now that the frame object has been defined and the appropriate positioning method in order, you can now start placing your widgets inside the window. Widgets can be specified directly in the class constructor for the specific application class but it is best to use a method called **create_widgets ()**. Using the create_widgets, you can now place the statement to create the widget. The constructor will look like the one below;

```
def __init__(self, master):
    super(Application, self).__init__(master)
self.grid()
self.create_widgets()
```

The create_widgets() method features all the statements necessary to build the widget objects that you desire in your window. The script1803.py program below is a good example of this;

```
 1: #!/usr/bin/python3
 2: from tkinter import *
 3:
 4: class Application(Frame):
 5:     """Build the basic window frame template"""
 6:
 7:     def __init__(self, master):
 8:         super(Application, self).__init__(master)
 9:         self.grid()
10:         self.create_widgets()
11:
12:     def create_widgets(self):
13:         self.label1 = Label(self, text='Welcome to my window!')
14:         self.label1.grid(row=0, column=0, sticky= W)
15:
16: root = Tk()
17: root.title('Test Application window with Label')
18: root.geometry('300x100')
19: app = Application(root)
20: app.mainloop()
```

Two lines of codes are used to specify a Label widget for the window in the create_widgets() method. The Line 13 above specifies the exact Label object, and line 14 uses the grid() method to position the Label widget in the window.

- **Defining the event handler for the window:** Another step involved in building a Graphical User Interface application is to specify the event that the window uses. Widgets that can generate events (such as when the application user clicks a button) use the command parameter to define the name of a method Python calls when it detects the event. For example, to link a button to an event method, you write code like this:

```
def create_widgets(self):
    self.button1 = Button(self, text="Submit", command = self.display)
    self.button1.grid(row=1, column=0, sticky = W)
def display(self):
    print("The button was clicked in the window")
```

The create_widgets() method will be able to create a single button to show in the window area. The Button class constructor makes the command parameter to be **self.display**, which points to the display () method in the class. For now, the test display () method will deploy a print () statement to show a message back in the command line, where you initiated the program.

The script **script1804.py** below will create a window with the button and the event specified;

```python
#!/usr/bin/python3
from tkinter import *
class Application(Frame):
    """Build the basic window frame template"""

    def __init__(self, master):
        super(Application, self).__init__(master)
        self.grid()
        self.create_widgets()

    def create_widgets(self):
        self.label1 = Label(self, text='Welcome to my window!')
        self.label1.grid(row=0, column=0, sticky=W)
        self.button1 = Button(self, text='Click me!', command=self.display)
        self.button1.grid(row=1, column=0, sticky=W)

    def display(self):
        """Event handler for the button"""
        print('The button in the window was clicked!')

root = Tk()
root.title('Test Button events')
root.geometry('300x100')
app = Application(root)
app.mainloop()
```

Exploring the tkinter widget

- **The Label widget:** you can put texts inside the editor with this widget. Use the text parameter to specify the text to display when you are adding a label widget to your window. See below;

  self.label1 = Label(self, text='This is a test label')

- **The button widget:** To let application users trigger an event that will allow them to read data in a form. The basic format for creating a button widget is shown below;

  self.button1 = Button(self, text='Submit', command=self.calculate)

- **The Checkbutton widget:** To make a choice between one or more items in a list.

- **Entry widget:** To create a single-line form field for displaying text dynamically in your window. See the format; *self.entry1 = Entry(self)*

- **The text widget:** Used to add a large amount of texts.

- **Listbox widget**: The Listbox widget gives a list with many values for your application user to select from. When you are using the Listbox widget, you can indicate how the user chooses items in the list with the **selectmode** parameter. See here;

  self.listbox1 = Listbox(self, selectmode=SINGLE)

The **selectmode** parameter has the following options;
- **SINGLE**—Which allows users to select only one item at a time.
- **BROWSE**— Which lets the user choose only one item but the items are movable within the list.
- **MULTIPLE**— Which lets the user choose multiple items by clicking them one after the other.
- **EXTENDED**— Which lets the user choose multiple items by using the Shift and Control keys while selecting items.

Once you have created the Listbox widget, you can add items to the list by using the insert () method as shown below;

self.listbox1.insert(END, 'Item One')

Python Web Programming
Installing a web server on your Raspberry pi using python

Running a web server on your pi

You can move your python application into the web world. But before you do that, you need to create a web server on the Raspberry pi to host these python applications. There are many web server applications that you can install inside the Raspberry pi. But the most popular one is the **Apache web server.** The **Apache web server** works well on the Raspberry pi as long as you don't try to host many users at the same time on the server concurrently.

Installing the Apache web server on your pi: The Raspbian OS contains many software in its repository, and the **Apache web server** is featured in a single software package called **Apache2.** To install, simply use the **apt – get** command. Enter the command below inside the LXTerminal;

pi@raspberypi ~ $ sudo apt-get install apache2

The Apache2 package will help you to install the Apache web server once you enter the above command. The supporting files needed to run the server will be installed along with the Apache web server. These supporting files include; **/var/www** (which is the folder for serving web documents), **/usr/lib/cgi-bin** (which is the folder for serving scripts), **/etc/apache2** (which serves as the folder for the web server configuration file) and **/var/log/error.log** (which is the Apache web server error log file).

Once the installation process is completed, the Apache web server will start automatically without necessarily using any manual method to get the server on the run. Nonetheless, you will be able to start and stop the Apache web server anytime you by using some basic command prompt. Use the command **sudo service apache2 stop** to stop the Apache web server. In the same vein, you can use the command **sudo service apache2 start** to start the server.

You can test the Apache web server as soon as you get it running. Open a browser from the LXDE desktop on the Raspberry pi, or you can connect to the Raspberry pi from another client on the network provided that you know the IP address of your Raspberry pi. If you want to connect from the pi's desktop, you can connect to the localhost name; *http://localhost/*

You can find the IP address assigned to your Raspberry pi by using the **ifconfig** command at the command prompt; *pi@raspberrypi ~ $ ifconfig*. Once you know the IP address assigned to the Raspberry pi, you can then go ahead to connect it from the remote client by typing the IP address as the URL. For instance, if you know your Raspberry pi IP address is 19.0.1.50, then you will use the URL *http://19.0.1.50/*.

CHAPTER SIX

Basic pi/python projects

Building your pi projects with python

Displaying HD image on the Raspberry pi using python script

HD images mean high-definition images. The dimension of an image will ordinarily tell you whether an image is in HD or not. The dimension of an image is usually measured in pixels and it is of width x height. If your picture has a dimension 1280x720; what this means is that the width of the picture is 1280 and the height is 720. Larger dimension images mean a high resolution and these kinds of images give a better picture when viewed. An HD image is of dimension 1280x720 pixel. You can check whether an image is HD by using the Image viewer utility that came default with the Raspbian OS. In the Raspbian Graphical User Interface on your Raspberry Pi, tap on the LXDE Programs Menu icon. Locate the file manager within the LXDE menu in the Accessories submenu. Upon opening your file manager, you should go to any image or photo you have previously saved on your Raspberry pi. When you see the photo or image in the file manager, right click on the photo to open it with the image viewer. The resolution of the image will be written clearly at the image viewer title bar together with the name of the image you opened.

Creating the image presentation script: There are many methods from the PyGame library that can be deployed to achieve this. You can import the PyGame library with the script below;

import pygame   *#Import PyGame library*

from pygame.locals import *   *#Load PyGame constants*

pygame.init()   *#Initialize PyGame*

Setting up the presentation screen: This is the stage where you are required to set the color you want as the background screen. You should consider the image you are working with while determining your background color. Essentially, you should deploy monochrome colors like gray, white, or black. You will be required to set the RGB color settings before you will be able to generate the background color. White color can be obtained by the RGB setting 255,255,255. The RGB setting for black color is 0, 0, and 0. The example below uses gray color with an RGB setting of 125, 125, and 125;

ScreenColor = Gray = 125,125,125

You need to be able to view the HD image on any screen size. This is only possible when you set the presentation screen in your script. With this, you don't need to modify your script to view your image on any screen size even if the computer is not yours. To get this done, initialize the screen using *pygame.display.set_mode* and you will be able to use the full screen as shown below;

ScreenFlag = FULLSCREEN | NOFRAME

PrezScreen = pygame.display.set_mode((0,0),ScreenFlag)

The above script makes the PyGame display screen to be set to be able to assume the full size of the screen the script is encountering at the particular moment. You can also use the NOFRAME flag in addition to using the FULL SCREEN. The NOFRAME will make any frame surrounding your screen to disappear during image presentation. This allows the entire screen to be available to display your HD image.

Finding your image: while finding your image, it is essential you know the file name. If you want your script to find your image for you, specify the location of the image and the image's file extension by using two variables. The example below has the variable **picturedirectory** set to tell you the exact location of the stored images;

PictureDirectory = '/home/pi/pictures'

PictureFileExtension = '.jpg'

The variable **PictureFileExtension** in the script above is telling you about the current file extension of the image. Once you have already specified the location of the image and the file extension, you can then list the files in the photo directory. You will need to import another module called the **os module** to get this done. See below to import the **os module;**

import os   #Import OS module

You can also use the **os.listdir** operation to list the contents of the directory that has the photos.
Each file spotted in the directory will be noticed by the Picture variable. Thus, you can then use a
"For loop" to process each file as it is being noticed;

for Picture in os.listdir(PictureDirectory):

If you have other files located in this same directory, grab and show only the desired image by having the script check each file for the appropriate file extension before you load it with the **.endswith** operation. You can deploy an "if statement" to carry out this task. The "for loop" you want to use to process each file before will now become;

for Picture in os.listdir(PictureDirectory):

   if Picture.endswith(PictureFileExtension):

With the scripts above, you will be able to process all the images that end with the appropriate file extension. The remaining part of the loop will only load each image and then fills your screen with set screen color;

Picture=PictureDirectory + '/' + Picture

```
Picture=pygame.image.load(Picture).convert_alpha()
PictureLocation=Picture.get_rect() #Current location
#Display HD Images to Screen ###############3
PrezScreen.fill(ScreenColor)
PrezScreen.blit(Picture,PictureLocation)
pygame.display.update()
```

Storing your images on a removable disc

The Raspbian operating system is loaded on to the Raspberry SD card. The OS has occupied some spaces, and you might sometimes need to store your images on a removable disc (like a flash drive) and rather not on the Pi SD card. This is to relieve the SD card of insufficient memory space issues. So, how then does the python script accesses the images you saved on the removable disc.? The Raspbian operating system deploys what we called a "device file" to access files (images) stored on a removable disc. It becomes important, then, to determine the "device file" name before you start working on the python script. Follow these steps to determine your device file name;
- Insert the removable disc (flash drive) into an appropriate port on the Raspberry pi. Open the drive in file manager by selecting "open in file manager" in the window displayed after successfully inserting the drive.

- Record the file directory on the address bar of the file manager window, and then close the file manager window. Let us assume that the directory name is /media/28BE-27DC when you checked.
- Open the LXTerminal and enter the command **ls directory.** In the command, replace the **directory** part with the file directory name (/media/28BE-27DC) that you recorded in the previous step, and then tap Enter. You will be able to access the pictures and any other files stored on the removable disc.
- Use the **mount** command to see the device file name that will be assigned to the removable disc when inserted into the Raspberry pi. See the below script to mount;

1: pi@raspberrypi ~ $ **mount**

2: /dev/root on / type ext4 (rw,noatime,data=ordered)

3: ...

4: /dev/mmcblk0p1 on /boot type vfat (rw,relatime,fmask=0022,dmask=0022,

5: codepage=cp437,iocharset=ascii,shortname=mixed,errors=remount-ro)

6: /dev/sda1 on /media/28BE-27DC type vfat (rw,nosuid,nodev,relatime,uid=1000,

7: ...
- The device file name is shown in line 6 and the name is */dev/sda1*. Record this name because you will need it in the python presentation script.
- After recording the device file name, you can then safely remove the removable device by typing *"umount device file name"* i.e *umount /dev/sda1*. Note the spelling of umount here and not unmount.
- Enter *mkdir /home/pi/pictures* to create a directory for the removable material (flash drive).

After determining the device file name for Raspbian to use to call your removable device, you can then make some adjustments to your presentation script. You need to, first, create a variable inside your python script to stand in for the device file name you have identified in the previous step. See below;

PictureDisk = '/dev/sda1'

The os module has a function referred to as **.system**. This **.system** function enables you to send dash shell commands from your Python script to the Raspbian operating system. You should ensure that the removable material has not been automatically mounted in your script. Simply issue the umount command by using the **os.system.** See below;

Command = "sudo umount " + PictureDisk

os.system(Command)

However, if your removable material has not been mounted, you will receive an error message for the prompt above and then proceed with the presentation, nonetheless. Having an error message here doesn't stop anything; only that error message won't make your script presentation look attractive. But this error message can be concealed by deploying a small dash command trick. See below;

Command = "sudo umount " + PictureDisk + " 2>/dev/null"

os.system(Command)

Now you can mount your removable material by using the Python script. Simply use the **os.system** function once again to issue a mount command to the operating system. See below;

Command = "sudo mount -t vfat " + PictureDisk + " " + PictureDirectory

os.system(Command)

Scaling your image: You don't want an oversized image that is much larger than your screen, so you need to scale your image. To do this, you need to first determine the size of the presentation screen by using the **.get_size** command. You can then scale down any image that is oversized by setting up a variable called **Scale.** See below to set up the **Scale** variable;

PrezScreenSize = PrezScreen.get_size()

Scale=PrezScreenSize

Prompt the "if statement" below to determine the size of the current picture;

If Picture is bigger than screen, scale it down.

if Picture.get_size() > PrezScreenSize:

    Picture = pygame.transform.scale(Picture,Scale)

Note that the image will be scaled down by using pygame.transform.scale above if it is bigger than the current presentation screen.

Framing the image: You need to frame the image to make it looks more appealing and attractive. A small adjustment to the **Scale** variable above will get the job done. See below;

Scale=PrezScreenSize[0]-20,PrezScreenSize[1]-20

Centering the image: This is necessary to place the image at the center of the presentation screen. The PyGame library, by default, displays your images at the upper left corner of the presentation screen. Use the variable called **CenterScreen** ton properly center your picture;

PrezScreenRect = PrezScreen.get_rect()

CenterScreen = PrezScreenRect.center

Your HD image will be looking more presentable having gone through the series of python scripting above.

The Sense HAT: Getting Raspberry to sense the world

The GPIO on the Raspberry pi will enable you to connect many external devices to the Raspberry pi. The Sense HAT allows your small computer (Raspberry) to communicate with the external environment.

You will learn how to control the LED matrix in the Sense HAT and how to collect sensor data.

In this project, you will need the following hardware and software;
- Raspberry pi and a Sense HAT (which are the hardware)
- Python v3 and Sense HAT for python v3 (these are the software and are already installed with the Raspbian operating system).

But if the Sense HAT package is not in your Raspberry pi, for one reason or another, enter the command *sudo apt-get install sense-hat* in the LXTerminal of the Raspberry to get it.

The Sense HAT is literally like an add-on board made specially for the Raspberry pi. This board (Sense HAT) will enable users to measure temperature, pressure, humidity and some other measurements. It can bring out information by using the built-in LED matrix.

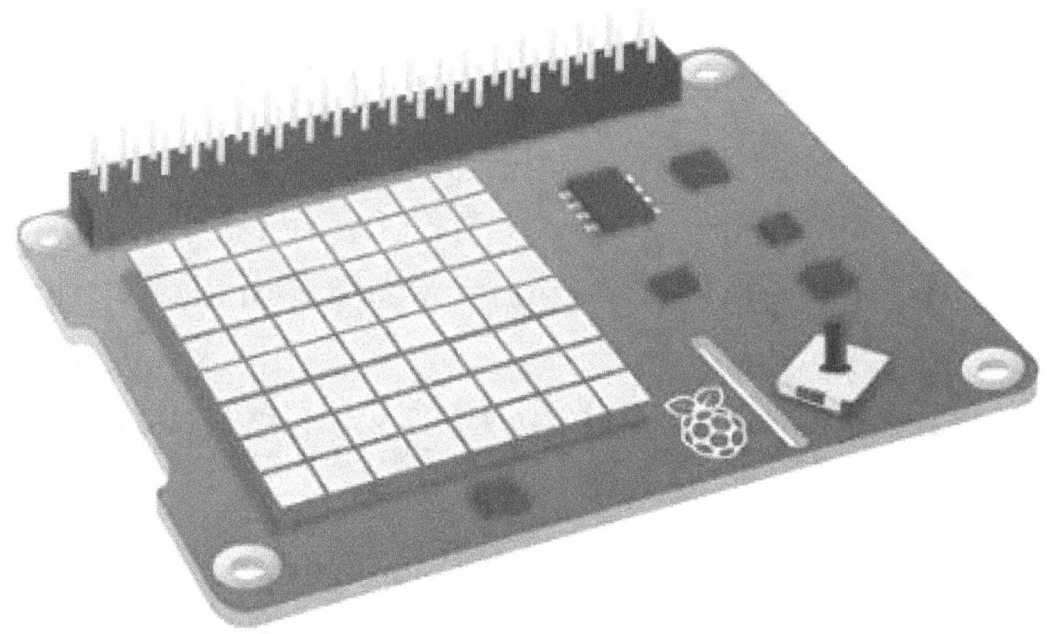

The Sense HAT can display text, color and images

- Prompt the code below to show a message on the LED matrix of the Sense HAT;

```
sense.show_message("Hello world")
```

- You can display color on the Sense HAT with a python script. Follow the steps;
    o Enter the script below in a python file.

```
from sense_hat import SenseHat

sense = SenseHat()

r = 255
g = 255
b = 255

sense.clear((r, g, b))
```

- o The LED matrix will show white light when you prompt the code above. This is because the maximum value of the red, green and blue colors has been used. Hence the color will be white.
- o Try something by changing the value of any of the colors above and note what you see.
- The LED matrix of the Sense HAT can be adjusted individually to show images
  - o The *set_pixels* command can be used to show images on the LED matrix.

Sensing the environment with the Sense HAT

There are some sets of sensors incorporated in the Sense HAT that enables it to make sense of the surroundings.

Reading pressure with the Sense HAT

Enter the code below in a python file

```
from sense_hat import SenseHat

sense = SenseHat()
sense.clear()

pressure = sense.get_pressure()
print(pressure)
```

Reading temperature with the Sense HAT

There are two sensors that are capable of reading the temperature. These include; the pressure sensor and the humidity sensor.

The *get_temperature_from_humidity* is capable of measuring the temperature from the humidity sensor, while the *get_temperature_from_pressure* can measure the temperature from the pressure sensor.

Open the python file and run the code the below;

from sense_hat import SenseHat

```
sense = SenseHat()
sense.clear()

temp = sense.get_temperature()
print(temp)
```

Reading humidity with the Sense HAT

Hit the code below in a python file;

```
from sense_hat import SenseHat

sense = SenseHat()
sense.clear()

humidity = sense.get_humidity()
print(humidity)
```

Sense HAT detecting movement

The Sense HAT has an Inertia Measurement Unit chip featuring some sensors that are capable of detecting motion. The sensors include;
- A gyroscope – which can detect which way up the board is going
- Accelerometer – which detects motion.
- Magnetometer- for magnetic field detection.

These movement sensors are important, particularly, in space navigation to know the way you are travelling in space. The IMU sensor can track movement in space.

ABOUT AUTHOR

Ted Humphrey is a tech expert and a programmer who understands python programming language and their application to 21st century problems. Ted has over 12 years of experience writing about latest gadgets and technical appliances. Ted is also a seasoned website developer with 7+ years of experience creating WordPress websites and experienced in using other content management systems, such as Drupal, aside WordPress to create websites. He values making an impact and has a blog where he teaches people the nitty-gritty of Programming Python, developing blogs using WordPress and other content management systems.

Ted holds a Bachelor's degree in Computer science and Engineering from the University of Michigan, USA. He loves pets and he is happily married with two beautiful daughters.

www.ingramcontent.com/pod-product-compliance
Lightning Source LLC
Chambersburg PA
CBHW060418220526
45465CB00008B/2926